Uwe Schilling

Measurements in Quantum Optics

Uwe Schilling

Measurements in Quantum Optics

Possibilities and Limitations

Südwestdeutscher Verlag für Hochschulschriften

Impressum/Imprint (nur für Deutschland/only for Germany)
Bibliografische Information der Deutschen Nationalbibliothek: Die Deutsche Nationalbibliothek verzeichnet diese Publikation in der Deutschen Nationalbibliografie; detaillierte bibliografische Daten sind im Internet über http://dnb.d-nb.de abrufbar.
Alle in diesem Buch genannten Marken und Produktnamen unterliegen warenzeichen-, marken- oder patentrechtlichem Schutz bzw. sind Warenzeichen oder eingetragene Warenzeichen der jeweiligen Inhaber. Die Wiedergabe von Marken, Produktnamen, Gebrauchsnamen, Handelsnamen, Warenbezeichnungen u.s.w. in diesem Werk berechtigt auch ohne besondere Kennzeichnung nicht zu der Annahme, dass solche Namen im Sinne der Warenzeichen- und Markenschutzgesetzgebung als frei zu betrachten wären und daher von jedermann benutzt werden dürften.

Verlag: Südwestdeutscher Verlag für Hochschulschriften GmbH & Co. KG
Dudweiler Landstr. 99, 66123 Saarbrücken, Deutschland
Telefon +49 681 37 20 271-1, Telefax +49 681 37 20 271-0
Email: info@svh-verlag.de

Approved by: Erlangen, Friedrich-Alexander-Universität, Diss., 2011

Herstellung in Deutschland:
Schaltungsdienst Lange o.H.G., Berlin
Books on Demand GmbH, Norderstedt
Reha GmbH, Saarbrücken
Amazon Distribution GmbH, Leipzig
ISBN: 978-3-8381-2980-8

Imprint (only for USA, GB)
Bibliographic information published by the Deutsche Nationalbibliothek: The Deutsche Nationalbibliothek lists this publication in the Deutsche Nationalbibliografie; detailed bibliographic data are available in the Internet at http://dnb.d-nb.de.
Any brand names and product names mentioned in this book are subject to trademark, brand or patent protection and are trademarks or registered trademarks of their respective holders. The use of brand names, product names, common names, trade names, product descriptions etc. even without a particular marking in this works is in no way to be construed to mean that such names may be regarded as unrestricted in respect of trademark and brand protection legislation and could thus be used by anyone.

Publisher: Südwestdeutscher Verlag für Hochschulschriften GmbH & Co. KG
Dudweiler Landstr. 99, 66123 Saarbrücken, Germany
Phone +49 681 37 20 271-1, Fax +49 681 37 20 271-0
Email: info@svh-verlag.de

Printed in the U.S.A.
Printed in the U.K. by (see last page)
ISBN: 978-3-8381-2980-8

Copyright © 2011 by the author and Südwestdeutscher Verlag für Hochschulschriften GmbH & Co. KG and licensors
All rights reserved. Saarbrücken 2011

Abstract

This thesis fathoms the possibilities and limitations of measurements in quantum optics. After a short introduction to the general aspects of measurements in quantum mechanics, it presents three examples of a measurement process in concrete applications to quantum optics.

For one, while operators describing quantities of interest might have a simple theoretical form, it is often hard to actually implement a measurement of their expectation values experimentally. This situation is encountered in Chapter 2 which is devoted to the measurement of quantum statistical properties of a light beam. In contrast to classical optics, these statistics are not fully determined by first- and second-order correlations. Instead, higher-order correlations, usually described by non-Hermitian operators, are necessary to characterize the quantum state completely. A particular example is the N-photon Fock state, for which only a measurement of the Nth-order correlations allows for a full characterization. Here, we present a new method how to measure an important class of Nth-order correlations for arbitrary N in a single spatial mode with two polarizations with limited experimental resources.

The second part is again centered around the realization of specific measurement operators for photon correlation measurements. However, in contrast to the previous chapter, these measurements are not an end in their own, but are designed to create specific quantum states as a consequence of the measurement process: A suitable measurement of photons emitted by a number of atoms allows to transfer the atoms into long-lived entangled states. We find that with this method it is possible to entangle two remote atomic qubits to an arbitrary and well-defined degree. Furthermore, we show how to generate two families of quantum states in an arbitrary number of remote atomic qubits: one family consists of all – symmetric and non-symmetric – total angular momentum eigenstates in N remote qubits, the other is a family of cluster states. It is also systematically investigated whether the technique of quantum state engineering by projective measurements allows for the creation of any arbitrary state. We find explicit solutions for the case of two and three qubits, while it is shown that no solution exist for $N \geq 7$. We conclude the section by proving that with some changes in the setup, this measurement-based quantum state engineering technique allows to create the same states among the scattered photons themselves rather than among the emitting atoms.

The last part is devoted to the amount of information one is able to gain about a given system when performing a measurement. The corresponding limits of information are investigated at the example of the duality that arises in a two-way interferometer between the visibility of the interference pattern and the knowledge about the path of the interfering object. We show that there exist correlations between the phase difference the interfering object acquires on its way through the interferometer and the amount of which-way information retrievable from certain observables of the which-way detector. In this way, the which-way information becomes a phase-dependent quantity. In particular, we find that for certain values of the phase shift, the amount of extractable which-way information can be larger than allowed by phase-independent measurement. This property is put to use to find the optimal observable for the which-way detector readout in dependence on the phase of the interfering object. Following this strategy, we are able to gain more which-way information than previously thought possible.

4

Contents

1 Introduction — 9

2 Correlation Measurements: Higher-Order Correlations in a Single Light Beam — 15
  - 2.1 Introduction — 15
  - 2.2 Correlation Functions — 16
  - 2.3 Setup for Measuring Correlations in a Single Light Beam — 18
  - 2.4 Measuring First- and Second-Order Correlations in a Single Light Beam — 19
    - 2.4.1 Measuring First-Order Correlation Functions — 19
    - 2.4.2 Measuring Second-Order Correlation Functions — 21
  - 2.5 Measuring Nth-Order Correlations — 22
    - 2.5.1 General Considerations — 22
    - 2.5.2 Finding Parameters for the Setup — 22
    - 2.5.3 Example — 26
  - 2.6 Conclusion — 27

3 Projective Measurements: Quantum State Engineering — 29
  - 3.1 Introduction — 29
  - 3.2 Tool Box — 30
    - 3.2.1 The Λ-level System — 30
    - 3.2.2 Atom-Photon Entanglement — 31
    - 3.2.3 Projective Measurements of Entangled Systems — 32
    - 3.2.4 Measuring Photons from more than one Atom — 33
    - 3.2.5 The Detection Operator — 34
  - 3.3 Heralded Entanglement of Arbitrary Degree in Remote Qubits — 36
    - 3.3.1 The Idea — 36
    - 3.3.2 Quantifying the Entanglement: A Malus' Law for the Concurrence — 38
    - 3.3.3 Experimental Feasibility — 41
    - 3.3.4 Conclusion — 43
  - 3.4 Generation of Total Angular Momentum Eigenstates in Remote Qubits — 44

		3.4.1	Introduction	44
		3.4.2	Total Angular Momentum Eigenstates	44
		3.4.3	Setup	45
		3.4.4	Measurement-based Preparation of Total Angular Momentum Eigenstates	45
		3.4.5	Examples	50
		3.4.6	Conclusion	51
	3.5	Generation of Cluster States in Remote Qubits		52
		3.5.1	Introduction	52
		3.5.2	A Proposal for Cluster States Generation by Xia *et al.*	52
		3.5.3	Translating the Proposal into a Fiber-based Setup	55
		3.5.4	Conclusion	58
	3.6	Generation of Arbitrary States: The Anystate?		59
		3.6.1	Fundamentals	59
		3.6.2	Physical Realization of an Arbitrary Detection Operator	60
		3.6.3	Generating the Anystate N=2	61
		3.6.4	Generating the Anystate for N=3	63
		3.6.5	Conclusion	64
	3.7	Interchanging the Roles of Atoms and Photons: A Versatile Source		66
		3.7.1	Introduction	66
		3.7.2	Setup	67
		3.7.3	Generation of Photonic Quantum States	69
		3.7.4	Conclusion	71

4 Complementary Measurements: Wave-Particle Duality   **73**

	4.1	Introduction		73
	4.2	Basic Concepts of Quantitative Wave-Particle Duality		75
		4.2.1	Quantitative Interference	75
		4.2.2	Quantitative Which-Way Knowledge	76
		4.2.3	Interferometer Formalisms	79
	4.3	Phase-dependent Wave-Particle Duality		83
		4.3.1	The Two-Atom interferometer: WW Information in First-Order Interference Effects	83
		4.3.2	The Two-Atom interferometer: WW Information in Second-Order Interference Effects	86
		4.3.3	General Interferometer	87
		4.3.4	Physical Interpretation of Phase-dependent WW information	89
	4.4	Improving the WW Knowledge by a Delayed Choice of the WWD Observable		91

	4.5	Further Improving the WW Knowledge	93
	4.6	Do We Still Obtain WW Information?	95
	4.7	Example: The Micromaser	96
	4.8	Conclusion	97
5	**Summary**		**99**
A	**Solving the Anystate Equations for $N=3$**		**101**
B	**The Natural Basis of an Arbitrary WWD**		**107**
	Bibliography		**111**

Chapter 1

Introduction

From the age of Galileo Galilei to the beginning of the 20th century, physicists have developed an intuitive understanding of the measurement process in what we now call classical physics: every quantifiable property was considered an objective reality, i.e., a well-defined value existed independently of whether a measurement was conducted or not. The measurement was to determine the value of this pre-existing quantity, and it could, in principle if not in practice, be conducted with an arbitrary accuracy and without changing the physical properties of the system under investigation. However, with the advent of an inherently non-deterministic theory, quantum mechanics, it was recognized that also the concept of a measurement had to be remodelled. For example, the theory presupposes that not every value of a measurable quantity is a valid experimental result. In this light, Albert Einstein correctly explained the photoelectric effect as a quantization of the electromagnetic field [1] and Otto Stern and Walther Gerlach showed experimentally that the angular momentum of a silver atom in its ground state along an arbitrary spatial direction was always found to be either $+\hbar/2$ or $-\hbar/2$, while it never took on any values in between [2]. From observations like these the measurement postulate of quantum mechanics was derived which is nowadays found in every textbook on quantum physics. We will follow Nielsen and Chuang to express it in a rather unconventional form which will be particularly useful for the purpose of this thesis [3]: A quantum measurement with a set of possible outcomes $\{m\}$ can be described by a collection of *measurement operators* $\{\hat{M}_m\}$ acting on the state space of the system being measured. If the state of the quantum system is $|\psi\rangle$ immediately before the measurement then the probability that result m occurs is given by

$$P(m) = \langle\psi|\hat{M}_m^\dagger \hat{M}_m|\psi\rangle. \tag{1.1}$$

As in any statistical theory, new information about the system in form of the measurement result has to be included whenever it is acquired, resulting in the so-called "collapse of the wavefunction"[1] which dictates that the final state $|\psi^{(f)}\rangle$ of the system after the measurement is given by

$$|\psi^{(f)}\rangle = \frac{\hat{M}_m|\psi\rangle}{\sqrt{\langle\psi|\hat{M}_m^\dagger \hat{M}_m|\psi\rangle}}. \tag{1.2}$$

In a loose manner of speaking, this process will be referred to as projection throughout the thesis, even though the operators $\hat{M}_m$ are not necessarily projectors in the mathematical sense. Thereby, the set of measurement operators satisfies the *completeness relation*

$$\sum_m \hat{M}_m^\dagger \hat{M}_m = \mathbb{1} \quad \Leftrightarrow \quad \sum_m \langle\psi|\hat{M}_m^\dagger \hat{M}_m|\psi\rangle = \sum_m P(m) = 1, \tag{1.3}$$

which is equivalent to the statement that probabilities for all possible outcomes sums up to one, i.e., that every measurement has a result. We note further the connection to the more familiar and important special case of *projective measurements*, where all elements of the collection $\{M_m\}$ are projectors, i.e. Hermitian and satisfying the condition $M_m M_{m'} = \delta_{m,m'} M_m$. In such a case, the measurement is described by the observable $\hat{W}$, a Hermitian operator which has the spectral decomposition

$$\hat{W} = \sum_m m M_m. \tag{1.4}$$

Thus, if the eigenvalues are non-degenerate, the set of projectors $\{\hat{M}_m\}$ can be written as $\{|W_m\rangle\langle W_m|\}$ where each element projects the wavefunction $|\psi\rangle$ on the eigenvectors $|W_m\rangle$ of the observable with eigenvalue m.

The present thesis does not attempt to give a comprehensive interpretation of the quantum mechanical measurement process itself or to solve corresponding open questions. Rather, it will accept the postulate as given and deal with questions that arise within this well-established framework in applications to quantum optics. In particular, three main problems will be investigated.

First, while it is simple to write down a set of measurement operators $\{M_m\}$ theoretically, it is often much harder to actually implement the measurement of their expectation values in an experiment. This situation is encountered in Chapter 2 which is devoted to the measurements of quantum statistical properties of a light beam. At optical frequencies, typical detection methods all reduce to photon counting, which means that only expectation values of intensity moments described by the operators $a_i^{\dagger N} a_i^N$ can be measured directly, where a_i and $a_i^\dagger$ denote the annihilation and creation operator of a photon in mode i, respectively. However, for a description of properties

---
[1] This process quite natural to a statistical theory but yet to be understood in its physical interpretation.

of the electromagnetic field like the polarization, also the expectation values of operators where the creation and annihilation operators for photons from different modes are mixed are of interest. A simple example of this would be the operator $\langle a_1^\dagger a_2 \rangle$, where the subscripts denote two orthogonal polarizations. In such a case, one first has to find Hermitian operators which will allow to derive the values of the quantities of interest and then find an experimental setup which allows to implement the measurement of these Hermitian operators. In the example chosen above, we will see that this could be, e.g., the Hermitian operators $a_1^\dagger a_2 + a_2^\dagger a_1$ and $-i(a_1^\dagger a_2 - a_2^\dagger a_1)$ which are implemented, not quite coincidentally, as simple polarization measurements. For the measurement of the expectation values in which products of more than one annihilation and one creation operator appear, the experimental difficulties rise quickly. However, for the characterization of quantum states, it is often indispensable to measure also these higher-order correlations because they are, e.g., an essential tool for a tomography of entangled photon states which occupy the same spatial mode [4, 5]. In Chapter 2, we will introduce a new measurement scheme which enables the measurement of arbitrary-order correlations in a single light beam in which equal powers of the creation and annihilation operator appear [6].

Chapter 3 is again centered around the implementation of specific measurement operators $\hat{M}_m$ and once more, they will be implemented as photon correlation measurements. However, in contrast to the previous chapter, these measurements are not an end in their own, but they are designed such as to create specific quantum states as a consequence of the measurement process. This is done by tailoring the measurement operators in such a way that, if successful, the projection occuring according to Eq. (1.2) casts the system into a certain desired state. Even from a classical point of view, it is not surprising that this is possible: if an experimenter has a statistical ensemble of systems in a completely random state, he will for sure find a subensemble in the specific state he had aimed for. However, in case of quantum state engineering, the eponymous property of quantum mechanics, namely that the measured quantities may be quantized and even have a finite number of allowed values, introduces a particular charm, because this property can augment the detection probability dramatically. Consider as a simple example an ensemble of spin-1/2 systems in which each system has a completely arbitrary direction of spin. If the spin could be considered as a classical angular momentum for which its component along an arbitrary axis $\mathbf{v}$ could assume any value, then it can be shown by simple geometrical considerations that only a fraction of $\theta^2/2$ of the spins would be found as having a spin direction which deviates by a small angle of less than θ from $\mathbf{v}$. The smaller θ, the smaller also the fraction of spins which will be found in the desired subensemble. However, it is well-known that the spin is quantized and consequently every measurement of a spin-1/2 system in a certain spatial direction $\mathbf{v}$ will always find the spin either parallel or antiparallel to that direction. Therefore, in an actual measurement, one will always find half of the systems in exactly the state that was aimed for.

The situation becomes even more interesting, if two systems A and B are entangled with each other and therefore show non-classical correlations. In such a case, a measurement of say system B will project the other system A into a certain state and therefore prepare that system remotely, i.e., without the need to control an interaction or the need to conduct a measurement, in a certain state. Moreover, the state of system A is *heralded* by the outcome of the measurement of system B, i.e., the state does not exist only post-selectively, as is often the case for example for photonic quantum states [7]. Bose *et al.* and Cabrillo *et al.* first proposed to use this technique to entangle two atoms which never interacted with each other by measuring the photons they scatter [8,9]. A similar approach is used in Chapter 3 where we explore the possibilities and limitations of this technique and explicitly show how to generate a large variety of quantum states from different families in massive remote qubits [10,11]. Furthermore, we demonstrate how these techniques can be translated to the generation also of photonic quantum states [12].

Finally, Chapter 4 is devoted to the limits of knowledge about the properties of a quantum mechanical process or system. It turns out that one consequence of the measurement postulate is that the information one is able to gain about a given system is limited in the sense that even if the state of a system is known perfectly, it is not possible to predict the outcome of any imaginable measurement with certainty. A spin-1/2 system may again serve as a simple example: assume we have prepared its spin to point along the positive z-axis, for example by the measurement procedure described above. In such a case, the state of the system is completely determined by its state vector $|\uparrow_z\rangle$ and we know that its component along the x-axis has to be 0. This, however, can only be true on average, since 0 is not a valid measurement result for a measurement of the spin along the x-axis. An actual single measurement will show with perfect randomness a result for the spin to be aligned either parallel or antiparallel to the x-axis. This is a manifestation of quantum statistics and at the same time a striking contradiction to the classical picture in which every measurable quantity of a system assumes a well-defined single value if the state of the system is completely known. In general, this phenomenon can be described by the discovery of Werner Heisenberg and Howard P. Robertson that if two operators $\hat{A}$ and $\hat{B}$, each describing a measurement, do not commute then it is not possible to assign precise values to both of them at the same time [13,14]. This is usually expressed in the form

$$\Delta\hat{A} \cdot \Delta\hat{B} \geq \frac{1}{2}\left|\langle[\hat{A},\hat{B}]\rangle\right|, \tag{1.5}$$

where $[\hat{A},\hat{B}] = \hat{A}\hat{B} - \hat{B}\hat{A}$ denotes the commutator. From this result, the famous Bohr-

Einstein debate emerged which was centered around the concept of interferometric duality [15]. In its simplest form, this concept states that in a two-way interferometer the acquisition of which-way information and the observation of an interference pattern are mutually exclusive. However, this statement only accounts for the extremal cases of obtaining no which-way information while observing a perfect interference pattern and vice versa. There exists a long and ongoing debate on the question how well one can measure the path of a quantum mechanical object in an interferometer when observing an interference pattern with a non-perfect visibility [16–19]. It started with a seminal paper by Wootters and Zurek in 1979 [16] who first quantified which-way information in the double-slit setup discussed by Bohr and Einstein. More recently, Englert derived a very general inequality for a generic two-way interferometer of the form $\mathcal{K}^2 + \mathcal{V}^2 \leq 1$ in which $\mathcal{K}$ quantifies the which-way information obtained from an auxiliary quantum system (the which-way detector) which measures the path of the interfering object and $\mathcal{V}$ represents the visibility of the interference pattern [18].

In the last part of the thesis, we probe this quantitative duality relation in a two-way interferometer discussed in Chapter 3: two atoms serve as emitters of one or two photons which are brought to interfere. We analyse in this system how much information can be gained about the path of the photon(s) by reading out the state of the atoms in dependence on the visibility of the interference pattern. It turns out that if the state of each atom is read out individually, the which-way information depends in addition on the detection position of the photon. This investigation paves the way to a very general statement, namely that in every two-way interferometer, there exists not only a correlation between the amount of obtainable which-way information and the visibility of the interference pattern, but there exist observables of the which-way detector for which also the phase difference that the interfering objects acquires on its path through the two arms of the interferometer is correlated with the amount of which-way information that is retrieved. In this way, the which-way information becomes a phase-dependent quantity [20]. It is particularly interesting to note that we can find observables of the which-way detector for which for certain values of the phase shift, the amount of extractable which-way information is larger than allowed by Englert's inequality. This property is put to use to find the optimal observable for the which-way detector readout in dependence on the phase shift of the interfering object. Following this strategy, we are all in all able to extract more which-way information from the which-way detector than previously thought possible [21].

Chapter 2

Correlation Measurements: Higher-Order Correlations in a Single Light Beam [6]

2.1 Introduction

Historically, the measurement of the properties of a light beam included the quantities direction, intensity, polarization, and wavelength (color). Nowadays, the classical picture is completed by a fifth quantity, the mutual coherence function [22]. However, for a complete quantal characterization of the field, it is necessary to measure so-called higher-order correlation functions (cf. Sec. 2.2). Unfortunately, these measurements are non-trivial. Only recently, the so-called covariance matrix of the Gaussian output beam of an optical parametric oscillator has been measured, which contains correlations of fourth order [23]. While this covariance matrix fully determines all characteristics of a Gaussian beam, it is not possible to verify the Gaussianity of the beam. In order to assure the Gaussian property, one needs to measure even higher-order correlation functions (in principal an unlimited number of orders).

A different reason for the interest in higher-order correlations is that in recent years interest in entangled states has grown significantly. One of the most successful experimental implementations has been achieved by entangling the polarization degrees of freedom of two photons in spontaneous parametric down conversion [24]. Since then, a lot of effort has been devoted to producing entangled states with an ever higher photon number [7, 25–28]. For the most part, entangled photon states are generated by postselection in such a way that every photon is found to have occupied an individual spatial mode [29]. In that kind of setup, a quantum state tomography is usually conducted with the help of quarter-wave plates and half-wave plates which operate on every output port separately in such a way that each mode is analyzed along different orthogonal bases.

The theory behind this method is intuitive and has been described exhaustively [30]. However, there exist experiments that generate polarization-entangled states of higher photon numbers in a single spatial mode [4, 5]. Here, it is possible to split the beam into as many spatial modes as there are photons to conduct a full state tomography as in [30], but for higher photon numbers, this approach is experimentally very costly and inefficient.

In this chapter, we discuss a method that does not require the beam to be separated into different spatial modes. It is based on a very general theorem, formulated by Mukunda and Jordan, which states that it is always possible to calculate the coherences (or correlations) of a photon field from photon correlation measurements in several different bases [31]. This theorem has been used in proposals to measure all second-order correlations of a field, corresponding to the variances of the so-called quantum Stokes parameters [32, 33]. Here, we show that it is possible to measure *all* Nth-order correlations in a light beam consisting of two polarization modes with an arbitrary and unknown amount of photons in each mode [6]. For this measurement we only require two quarter-wave plates, one half-wave plate, and a polarizing beam splitter, all acting on the same single spatial mode. Then, the Nth-order intensity moment of that mode is measured after passage of the photons through the mentioned optical elements. If the incoming field is in a Fock state of N photons, this procedure corresponds to a full state tomography of that state (cf. Sec. 2.5.1). We note that the scheme is not limited to polarization optics, but may also be applied to other two-mode systems where the necessary operations can be implemented, for example photons in two different Laguerre-Gaussian modes [34–36].

2.2 Correlation Functions

Let us start with a short recapitulation of correlations of the electromagnetic field. The electromagnetic field operator $\hat{\mathbf{E}}(\mathbf{r}, t)$, quantized in a volume L^3, is given by [22]

$$\hat{\mathbf{E}}(\mathbf{r},t) = \frac{1}{L^{3/2}} \sum_{\mathbf{k},s} \left(\frac{\hbar\omega}{2\varepsilon_0} \right)^{1/2} \left[i\boldsymbol{\varepsilon}_{\mathbf{k}s} e^{i(\mathbf{k}\cdot\mathbf{r}-\omega t)} \hat{a}_{\mathbf{k}s} - i\boldsymbol{\varepsilon}_{\mathbf{k}s}^* e^{-i(\mathbf{k}\cdot\mathbf{r}-\omega t)} \hat{a}_{\mathbf{k}s}^\dagger \right]. \quad (2.1)$$

Here, ε_0 is the dielectric constant of the vacuum, $\boldsymbol{\varepsilon}_{\mathbf{k}s}$ denotes the (unit) polarization vector of the s-component of the field in mode $\mathbf{k}$ and frequency $\omega = \omega(\mathbf{k}) = c|\mathbf{k}|$ where s is one of two orthogonal polarization orientations and $\hat{a}_{\mathbf{k}s}$ and $\hat{a}_{\mathbf{k}s}^\dagger$ are the annihilation and creation operator of a photon in mode $\mathbf{k}$ and with polarization s, respectively, which obey the conventional bosonic commutation relation $[a_{\mathbf{k}s}, a_{\mathbf{k}'s'}^\dagger] = \delta_{\mathbf{k}\mathbf{k}'} \delta_{ss'}$. It is often convenient to rearrange the sum and separate the positive and negative frequency

2.2. CORRELATION FUNCTIONS

parts of the operator

$$\hat{\mathbf{E}}(\mathbf{r},t) = \underbrace{A_0 e^{-i\omega t} \sum_{\mathbf{k},s} \boldsymbol{\varepsilon}_{\mathbf{k}s} e^{i\mathbf{k}\cdot\mathbf{r}} \hat{a}_{\mathbf{k}s}}_{\hat{\mathbf{E}}^+(\mathbf{r},t)} + \underbrace{A_0^* e^{i\omega t} \sum_{\mathbf{k},s} \boldsymbol{\varepsilon}_{\mathbf{k}s}^* e^{-i\mathbf{k}\cdot\mathbf{r}} \hat{a}_{\mathbf{k}s}^\dagger}_{\hat{\mathbf{E}}^-(\mathbf{r},t)}, \qquad (2.2)$$

where $A_0 = \left(\frac{-\hbar\omega}{2\varepsilon_0 L^3}\right)^{1/2}$ and we have assumed that we are working with quasi-monochromatic light ($\omega = $ const.).

The electric field $E(\mathbf{x})$ displays a characteristic statistical distribution around its mean $\langle E(\mathbf{x})\rangle$ at every point $\mathbf{x} = (\mathbf{r},t)$ in space-time. However, the properties of the electric field are not fully determined by its statistical distribution at every point $\mathbf{x}$, because this specification does not contain any information about *correlations* between the value of the electric field vector at two (or more) points in space time. The usually most important correlations are described by the coherence matrix consisting of the four mutual coherence functions $\Gamma_{ij}^{(1,1)}(\mathbf{x}_1,\mathbf{x}_2)$ which are defined as

$$\Gamma_{ij}^{(1,1)}(\mathbf{x}_1,\mathbf{x}_2) = \langle \hat{\mathbf{E}}_i^-(\mathbf{x}_1)\hat{\mathbf{E}}_j^+(\mathbf{x}_2)\rangle, \qquad (2.3)$$

where the subscripts i and j refer to one of the two orthogonal polarizations $s_{1,2}$. If a field obeys Gaussian statistics, all higher-order correlation functions can be decomposed via the moment theorem into a sum of products of the mutual coherence functions[1], and thus, classical fields are completely characterized by their mutual coherence functions [22].

However, a quantum field does not necessarily obey Gaussian statistics and therefore the mentioned quantities do not capture the quantum nature of light, i.e., light might have properties which remain hidden when measuring only its mutual coherence function. For a full characterization of the quantum field, one therefore needs to know the correlations of the field $\Gamma^{M,N}$ for all orders $M, N \in \mathbb{N}$ at positions $\mathbf{x}_1, \ldots, \mathbf{x}_{M+N}$, and all polarizations $i_1, i_2, \ldots, i_M, j_1, j_2, \ldots, j_N$, i.e., all correlation functions of the form

$$\Gamma_{i_1,i_2,\ldots,i_M,j_1,j_2,\ldots,j_N}^{(M,N)} =$$
$$\langle \hat{\mathbf{E}}_{i_1}^-(\mathbf{x}_1)\hat{\mathbf{E}}_{i_2}^-(\mathbf{x}_2)\ldots\hat{\mathbf{E}}_{i_M}^-(\mathbf{x}_M)\hat{\mathbf{E}}_{j_1}^+(\mathbf{x}_{M+1})\hat{\mathbf{E}}_{j_2}^+(\mathbf{x}_{M+2})\ldots\hat{\mathbf{E}}_{j_N}^+(\mathbf{x}_{M+N})\rangle. \qquad (2.4)$$

In principle, measuring these expectation values for all values of M and N will allow one to characterize the state of the field completely. However, in practice, one is usually content with the measurements of certain correlations only. Thus, for the remainder of

---

[1] This is not surprising since a Gaussian is fully defined by its mean and its variance.

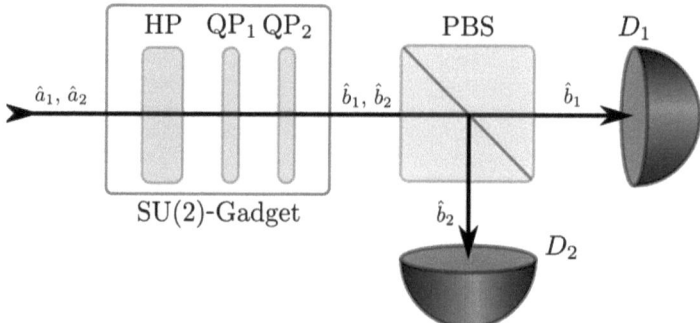

Figure 2.1: A sketch of the setup with two quarter-wave plates (QP$_1$ and QP$_2$), one half-wave plate (HP), and a polarizing beam splitter (PBS). At the detectors ($D_{1,2}$), the Nth-order intensity moment is measured.

the thesis, only correlation functions of *even* order $\Gamma^{(N,N)} = \Gamma^{(N)}$ will be considered, because they do not require a local oscillator and are therefore particularly easy to measure. In addition, as we will see, they suffice for a full state tomography of a Fock state. In a somewhat loose wording, they will also be called Nth-order correlations, because they can be considered to be Nth order in the intensity.

2.3 Setup for Measuring Correlations in a Single Light Beam

As outlined in the introduction, we will discuss the correlations of a single light beam only. For the measurement of these correlations, the setup shown in Fig. 2.1 will be used. The light beam first encounters a half-wave plate and two quarter-wave plates, all mounted coaxially. The action of the half-wave plate on the two modes $(\hat{a}_1, \hat{a}_2)$ of orthogonal linear polarization in the single beam is given by

$$H(\alpha) = i \begin{pmatrix} \cos(2\alpha) & \sin(2\alpha) \\ \sin(2\alpha) & -\cos(2\alpha) \end{pmatrix} \tag{2.5}$$

and that of the quarter-wave plates by

$$Q(\alpha) = \frac{i}{\sqrt{2}} \begin{pmatrix} \cos(2\alpha) - i & \sin(2\alpha) \\ \sin(2\alpha) & -\cos(2\alpha) - i \end{pmatrix}, \tag{2.6}$$

where α is the angle between the slow axis of the wave plates and the plane defined by the polarization of $\hat{a}_1$. Simon and Mukunda showed that the wave plate arrangement

2.4. MEASURING SIMPLE CORRELATION FUNCTIONS

described above constitutes a universal SU(2) gadget for polarized light and implements a general rotation in SU(2) space [37]. Thus, the action of this SU(2) gadget on the two modes $\hat{a}_1$ and $\hat{a}_2$ imposes a transformation parametrized by the two angles θ and ϕ

$$\begin{pmatrix}\hat{b}_1\\\hat{b}_2\end{pmatrix} = U(\theta,\phi)\begin{pmatrix}\hat{a}_1\\\hat{a}_2\end{pmatrix} \qquad (2.7)$$

with

$$U(\theta,\phi) = Q(\alpha_{\mathrm{QP}_2})Q(\alpha_{\mathrm{QP}_1})H(\alpha_{\mathrm{HP}}) = \begin{pmatrix}\cos\theta & e^{i\phi}\sin\theta\\-e^{-i\phi}\sin\theta & \cos\theta\end{pmatrix}, \qquad (2.8)$$

where θ and ϕ are abstract parameters determined by the orientation of the three wave plates; the exact functional dependence is given in Sec. 2.5.3. After the wave plates, a polarizing beam splitter seperates the modes $\hat{b}_1$ and $\hat{b}_2$ which are then individually recorded by a detector. The detectors are assumed to be capable of measuring Nth order intensity moments I^N which where shown by Glauber to be proportional to [38]

$$I^N \propto \langle \hat{b}_X^{\dagger N} \hat{b}_X^N \rangle. \qquad (2.9)$$

For small N, this can be realized, e.g., by splitting each mode into two or more parts and measuring and correlating the intensities at the output ports. Better scalability to higher values of N is provided if the the beam diameter is extended with a telescope setup and the intensity correlations are measured with a CCD chip (e.g. [39]).

2.4 Measuring First- and Second-Order Correlations in a Single Light Beam

2.4.1 Measuring First-Order Correlation Functions

An often-used description for the polarization state of light are the Stokes parameters, which describe the state of a polarized light beam as a point of the Poincaré sphere (cf. Fig. 2.2). For a classical coherent beam, the parameters are defined by:

$$\begin{aligned}S_0 &= |\alpha_1|^2 + |\alpha_2|^2, & S_1 &= |\alpha_1|^2 - |\alpha_2|^2,\\ S_2 &= \alpha_1^*\alpha_2 + \alpha_1\alpha_2^*, & S_3 &= -i(\alpha_1^*\alpha_2 - \alpha_1\alpha_2^*),\end{aligned} \qquad (2.10)$$

where α_1 and α_2 represent the amplitude of the beam in two orthogonal linear polarizations. Since in the present chapter, we are interested in measuring the correlations in a single light beam, i.e., in a single spatial mode, the summation over $\mathbf{k}$ in the quantum description Eq. (2.2) drops out. Under these additional assumptions (and dropping

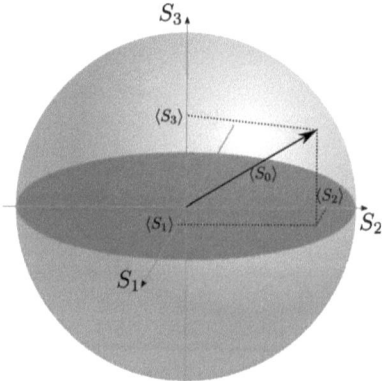

Figure 2.2: Representation of the Stokes parameters with the Poincaré sphere. The intensity S_0 of the beam defines the radius of the sphere, S_1 corresponds to its linear polarization along the axes of $s_{1,2}$, S_2 to the linear polarization under $\pm 45°$ with respect to those axes, and S_3 denotes its circular polarization. All vectors representing completely polarized light lie on the surface of the Poincaré sphere while those for partially polarized light lie within.

proportionality constants), one can identify for ease of notation the positive and the negative frequency part of the field directly with the photon annihilation and creation operator, respectively. Therefore, the quantum Stokes parameters are derived from the classical Stokes parameters by replacing the amplitudes and their complex conjugates with the annihilation and creation operators of the electric field components. This leads to the following definition for the quantum Stokes parameters [40, 41]:

$$\begin{aligned} \hat{S}_0 &= \hat{a}_1^\dagger \hat{a}_1 + \hat{a}_2^\dagger \hat{a}_2, & \hat{S}_1 &= \hat{a}_1^\dagger \hat{a}_1 - \hat{a}_2^\dagger \hat{a}_2, \\ \hat{S}_2 &= \hat{a}_1^\dagger \hat{a}_2 + \hat{a}_2^\dagger \hat{a}_1, & \hat{S}_3 &= -i(\hat{a}_1^\dagger \hat{a}_2 - \hat{a}_2^\dagger \hat{a}_1). \end{aligned} \quad (2.11)$$

From the measurement of the expectation values $\langle S_i \rangle$ of the quantum Stokes parameters, the four averages $\langle \hat{a}_1^\dagger \hat{a}_1 \rangle$, $\langle \hat{a}_1^\dagger \hat{a}_2 \rangle$, $\langle \hat{a}_2^\dagger \hat{a}_1 \rangle$, and $\langle \hat{a}_2^\dagger \hat{a}_2 \rangle$ can be deduced by inverting the system of equations (2.11). Schemes to measure these four quantities are standard textbook material, since the procedure is equivalent to determining the polarization of a light beam and their values equal those of their classical counterparts [42].

In our setup (cf. Sec. 2.3), a measurement of the intensity $\langle \hat{b}_1^\dagger \hat{b}_1 \rangle_{(\theta,\phi)}$ in mode $\hat{b}_1$, after the waveplates have induced a transformation $U(\theta, \phi)$, is given by

$$\begin{aligned} \langle \hat{b}_1^\dagger \hat{b}_1 \rangle_{(\theta,\phi)} &= \langle (U(\theta,\phi)\hat{\mathbf{a}})_1^\dagger (U(\theta,\phi)\hat{\mathbf{a}})_1 \rangle = \\ &\cos^2\theta \langle \hat{a}_1^\dagger \hat{a}_1 \rangle + \sin^2\theta \langle \hat{a}_2^\dagger \hat{a}_2 \rangle + e^{i\phi}\sin\theta\cos\theta \langle \hat{a}_1^\dagger \hat{a}_2 \rangle + e^{-i\phi}\sin\theta\cos\theta \langle \hat{a}_2^\dagger \hat{a}_1 \rangle. \quad (2.12) \end{aligned}$$

2.4. MEASURING SIMPLE CORRELATION FUNCTIONS

By setting the waveplates such that the parameters (θ, ϕ) subsequently assume the values $(0,0)$, $(\pi/2, 0)$, $(\pi/4, 0)$, and $(\pi/4, \pi/2)$, the measurements described in [42] to obtain the averages $\langle S_i \rangle$ are exactly reproduced. However, any other set of parameters $\{(\theta_i, \phi_i), i = 1, 2, 3, 4\}$ for which one obtains four independent linear equations from Eq. (2.12) would allow to deduce the expectation values of the quantum Stokes parameters as well.

2.4.2 Measuring Second-Order Correlation Functions

While the expectation values of the quantum Stokes parameters assume the same values as their classical counterparts, the expectation values of their variances, defined by the difference of two anticommutators

$$V_{ij} = \frac{1}{2}(\langle \{S_i, S_j\} \rangle - \{\langle S_i \rangle, \langle S_j \rangle\}), \tag{2.13}$$

with $i, j = \{0, 1, 2, 3\}$, are not equal to their classical counterparts. For example, in the classical picture all variances of a coherent beam should vanish, whereas the analysis of the quantized fields shows that the commutators for the quantum Stokes parameters are given by

$$\begin{aligned}[S_i, S_j] &= 2i\epsilon_{ijk} S_k \qquad i, j, k \in \{1, 2, 3\}, \\ [S_0, S_i] &= 0.\end{aligned} \tag{2.14}$$

Consequently, the quantum Stokes parameters obey an operator-valued uncertainty relation of the Heisenberg-Robertson kind (cf. Eq. (1.5)) and, thus, their variances cannot all vanish at the same time [43]. The variances can be described in terms of all nine normally ordered second-order field correlations: $\langle \hat{a}_1^\dagger \hat{a}_1^\dagger \hat{a}_1 \hat{a}_1 \rangle$, $\langle \hat{a}_1^\dagger \hat{a}_1^\dagger \hat{a}_1 \hat{a}_2 \rangle$, $\langle \hat{a}_1^\dagger \hat{a}_2^\dagger \hat{a}_1 \hat{a}_1 \rangle$, $\langle \hat{a}_1^\dagger \hat{a}_2^\dagger \hat{a}_1 \hat{a}_2 \rangle$, $\langle \hat{a}_1^\dagger \hat{a}_1^\dagger \hat{a}_2 \hat{a}_2 \rangle$, $\langle \hat{a}_1^\dagger \hat{a}_2^\dagger \hat{a}_2 \hat{a}_2 \rangle$, $\langle \hat{a}_2^\dagger \hat{a}_2^\dagger \hat{a}_1 \hat{a}_2 \rangle$, $\langle \hat{a}_2^\dagger \hat{a}_2^\dagger \hat{a}_1 \hat{a}_1 \rangle$, and $\langle \hat{a}_2^\dagger \hat{a}_2^\dagger \hat{a}_2 \hat{a}_2 \rangle$. Out of those, only the three field correlations $\langle \hat{a}_1^\dagger \hat{a}_1^\dagger \hat{a}_1 \hat{a}_1 \rangle$, $\langle \hat{a}_1^\dagger \hat{a}_2^\dagger \hat{a}_1 \hat{a}_2 \rangle$, and $\langle \hat{a}_2^\dagger \hat{a}_2^\dagger \hat{a}_2 \hat{a}_2 \rangle$, which correspond to second-order intensity moment measurements, are directly accessible in experiments.

Korolkova *et al.* first proposed a way to measure the diagonal variances $\langle V_{ii} \rangle$ by using a half-wave plate and a quarter-wave plate [43] and Bowen *et al.* presented an experimental implementation [44]. Later, Agarwal and Chaturvedi showed that it is possible to measure all variances V_{ij} of the quantum Stokes parameters (i.e., all second-order correlations of light) by conducting measurements of intensity-intensity correlations in a setup as descriped above (cf. Sec. 2.3). Their approach was to measure $\langle \hat{b}_1^{\dagger 2} \hat{b}_1^2 \rangle_{(\theta, 0)}$ for five different values of θ, $\langle \hat{b}_1^{\dagger 2} \hat{b}_1^2 \rangle_{(\theta, \pi/2)}$ for three different values of θ and $\langle \hat{b}_1^\dagger \hat{b}_2^\dagger \hat{b}_1 \hat{b}_2 \rangle_{(\pi/4, \pi/4)}$. For these parameters, the measured quantities in terms of the nine normally ordered second-order field correlations formed an invertible system of linear equations and thus allowed also allowed for the calculation of all V_{ij} [32].

2.5 Measuring Nth-Order Correlations

2.5.1 General Considerations

The knowledge of $\langle \hat{\mathbf{S}} \rangle$ and $\langle \hat{V}_{ij} \rangle$ already gives a good idea of the nature of the quantum state; however, the measurement of higher-order correlations may add even more information about that state, in some cases it is even mandatory to measure them to fully describe the state of the field. In particular, if the measured beam is in a photon-number state with N photons, that is, if the photonic state is of the form

$$\sum_{n=0}^{N} c_n |n\rangle_1 |N-n\rangle_2, \qquad \sum_{n=0}^{N} |c_n|^2 = 1, \qquad (2.15)$$

the density matrix of that state has a size of $(N+1) \times (N+1)$ and its $(N+1)^2$ elements correspond to all Nth-order correlations. Thus, for an N-photon Fock state, the measurement of all Nth-order correlations is equivalent to a full state tomography.

In the following, we generalize the approach taken in Sec. 2.4 and show that with the setup depicted in Fig. 2.1, it is possible to determine *all Nth-order correlations* for arbitrary N by measuring only Nth-order *intensity moments*.

Using the parametrization of the unitary transformation given in Eq. (2.7), we can express the most general case of measuring the Nth-order correlation – defined by the correlation of the ith intensity moment in mode $\hat{b}_1$ with the $(N-i)$th intensity moment in mode $\hat{b}_2$ – behind the SU(2) gadget as

$$\langle \hat{b}_1^{\dagger i} \hat{b}_2^{\dagger N-i} \hat{b}_1^{i} \hat{b}_2^{N-i} \rangle = \sum_{w,y=0}^{i} \sum_{x,z=0}^{N-i} \binom{i}{w}\binom{i}{y}\binom{N-i}{x}\binom{N-i}{z} (\cos\theta)^{2N-w-x-y-z}$$
$$\times (\sin\theta)^{w+x+y+z} (-1)^{x+z} e^{i\phi(x+y-w-z)} \langle \hat{a}_1^{\dagger i+x-w} \hat{a}_2^{\dagger N-i-x+w} \hat{a}_1^{i+z-y} \hat{a}_2^{N-i-z+y} \rangle. \qquad (2.16)$$

To solve for the $(N+1)^2$ independent real variables in this equation, we must perfom at least $(N+1)^2$ measurements. Hereby, we must be sure that the values of θ and ϕ chosen for these measurements lead to a system of independent linear equations; from Eq. (2.16), it is not obvious that this is possible for arbitrary N. In the following, a set of values for θ and ϕ is given for which we show that the measurement of Nth-order intensity moments leads to a solvable system of equations. In the course of this proof, a natural recipe is developed that describes how the measurement results can be easily related to the correlations of the initial state.

2.5.2 Finding Parameters for the Setup

The results of this section show that it suffices to measure photons of just one polarization, either $\hat{b}_1$ or $\hat{b}_2$, to determine all correlations. For this reason, it is enough to set

2.5. MEASURING NTH-ORDER CORRELATIONS

up a measurement apparatus behind only one of the two output ports of the polarizing beam splitter (cf. Fig. 2.1). In a random choice, we opt to measure the Nth-order intensity in mode $\hat{b}_1$ (i.e. $i = N$) and can therefore drop the summation over x and z in Eq. (2.16), which consequently simplifies to

$$\langle \hat{b}_1^{\dagger N} \hat{b}_1^N \rangle = \sum_{w,y=0}^{N} \binom{N}{w}\binom{N}{y} (\cos\theta)^{2N-w-y} (\sin\theta)^{w+y} e^{i\phi(y-w)} \langle \hat{a}_1^{\dagger N-w} \hat{a}_2^{\dagger w} \hat{a}_1^{N-y} \hat{a}_2^{y} \rangle. \tag{2.17}$$

Experimentally, $\langle \hat{b}_1^{\dagger N} \hat{b}_1^N \rangle$ corresponds to a measurement of the Nth-order intensity moment. Note that the number of terms in Eq. (2.17) is $(N+1)^2$ and the expectation value of each correlation and population appears exactly once. Thus, the system of linear equations generated from this equation by $(N+1)^2$ measurements of $\langle \hat{b}_1^{\dagger N} \hat{b}_1^N \rangle$ for different ϕ and θ has exactly one solution if and only if we can choose the values of every pair (ϕ, θ) such that all equations are independent. To arrive at such a choice, we first introduce new indices of summation, α and β, such that we can rewrite Eq. (2.17) in a form where the phase $e^{i\beta\phi}$ factors out of one sum:

$$\langle \hat{b}_1^{\dagger N} \hat{b}_1^N \rangle = \sum_{\beta=-N}^{N} e^{i\beta\phi} \sum_{\alpha \in G_\beta} \binom{N}{\frac{\alpha+\beta}{2}} \binom{N}{\frac{\alpha-\beta}{2}} (\cos\theta)^{2N-\alpha}$$
$$(\sin\theta)^\alpha \langle \hat{a}_1^{\dagger N-\frac{\alpha-\beta}{2}} \hat{a}_2^{\dagger \frac{\alpha-\beta}{2}} \hat{a}_1^{N-\frac{\alpha+\beta}{2}} \hat{a}_2^{\frac{\alpha+\beta}{2}} \rangle, \tag{2.18}$$

with

$$\alpha = y + w, \qquad \beta = y - w,$$

and

$$G_\beta = \{2(N-\kappa) - |\beta|\} \text{ with } \kappa \in \{0, 1, \ldots, N - |\beta|\}. \tag{2.19}$$

Equation (2.18) is the starting point for our analysis. Please note that for all $k-1$ kth roots of unity r_l (except unity itself), the equation $\sum_{\kappa=0}^{k-1} (r_l)^\kappa = 0$ holds.[2] This useful identity is exploited by choosing ϕ adequately to simplify Eq. (2.18) further and to introduce a proof by example which shows that for suitable choices of ϕ and θ, the system of linear equations resulting from Eq. (2.17) can be solved. However, in order to arrive at a complete solution, we must distinguish in the following between measuring

---

[2] This can be seen instantaneously from the formula for the geometric series:

$$\sum_{i=0}^{k-1} q^i = \frac{1-q^k}{1-q}.$$

correlations of an odd or an even order N.

N even

If N is even, we choose for ϕ the values $\phi_k = \frac{2\pi k}{N+1}$ with $k \in \{1, 2, \ldots, N+1\}$. Consequently, $e^{\pm i\phi_k}$ corresponds to all $(N+1)$th roots of unity. For every choice of ϕ, we perform a measurement for $N+1$ different values of θ, with $\theta_j = \frac{j}{N+2}\frac{\pi}{2}$ and $j \in \{1, 2, \ldots, N+1\}$, thus carrying out $(N+1)^2$ measurements and obtaining $(N+1)^2$ different equations from Eq. (2.18). By summing all equations of equal θ_j, all terms from Eq. (2.18) with $\beta \neq 0$ cancel because of the mentioned property of the roots of unity, and the sum over β contracts to $\beta = 0$. Thus, we are left with $N+1$ equations (one for every value of j) containing only the $N+1$ diagonal terms, each depending on a *different* power of $\cos\theta_j$:

$$\frac{1}{N+1} x_{\theta_j} = \sum_{\alpha \in G_\beta} \binom{N}{\frac{\alpha}{2}} \binom{N}{\frac{\alpha}{2}} (\cos\theta_j)^{2N-\alpha} (\sin\theta_j)^\alpha \langle \hat{a}_1^{\dagger N-\frac{\alpha}{2}} \hat{a}_2^{\dagger \frac{\alpha}{2}} \hat{a}_1^{N-\frac{\alpha}{2}} \hat{a}_2^{\frac{\alpha}{2}} \rangle, \qquad (2.20)$$

where $x_{\theta_j} = \sum_{\phi_k} x_{\theta_j}^{\phi_k}$ and $x_{\theta_j}^{\phi_k}$ is the result of the N photon measurement $\langle \hat{b}_1^{\dagger N} \hat{b}_1^N \rangle$ for setting $\phi = \phi_k$ and $\theta = \theta_j$. This set of equations can now be solved for the diagonal terms.

We need not make more measurements to determine the other correlations. By first multiplying Eq. (2.18) by $e^{i\phi_k}$, and then adding all measurements for identical θ_j, only terms with $\beta = -k$ and $\beta' = N - k + 1$ survive and we arrive at

$$\frac{e^{i\phi_k}}{N+1} x_{\theta_j} = \sum_{\alpha \in G_\beta} \binom{N}{\frac{\alpha+\beta}{2}} \binom{N}{\frac{\alpha-\beta}{2}} (\cos\theta_j)^{2N-\alpha} (\sin\theta_j)^\alpha \langle \hat{a}_1^{\dagger N-\frac{\alpha-\beta}{2}} \hat{a}_2^{\dagger \frac{\alpha-\beta}{2}} \hat{a}_1^{N-\frac{\alpha+\beta}{2}} \hat{a}_2^{\frac{\alpha+\beta}{2}} \rangle$$
$$+ \sum_{\alpha' \in G_{\beta'}} \binom{N}{\frac{\alpha'+\beta'}{2}} \binom{N}{\frac{\alpha'-\beta'}{2}} (\cos\theta_j)^{2N-\alpha'} (\sin\theta_j)^{\alpha'} \langle \hat{a}_1^{\dagger N-\frac{\alpha'-\beta'}{2}} \hat{a}_2^{\dagger \frac{\alpha'-\beta'}{2}} \hat{a}_1^{N-\frac{\alpha'+\beta'}{2}} \hat{a}_2^{\frac{\alpha'+\beta'}{2}} \rangle.$$

(2.21)

Since N is even, all α are odd, while all α' are even or vice versa (cf. Eq. (2.19)), leaving a total sum, in which each correlation term again depends on a different power of $\cos\theta_j$.[3] Furthermore, the total number of correlation terms appearing in Eq. (2.21) is given by $|G_\beta| + |G_{\beta'}|$ which is equal to $N+1$ for every choice of k. Thus, the system of linear equations generated from Eq. (2.21) by inserting all $N+1$ values of θ_j is solvable. Furthermore, ϕ_k determines the correlations that appear in the system. It is enough to generate a system of linear equations for every ϕ_k with $k \in \{1, 2, \ldots, N/2\}$ to solve for all correlations. Since the total number of measurements is equal to the

---
[3] Remember that each correlation term appears only once in Eq. (2.17) and thus also maximally once in Eq. (2.21).

2.5. MEASURING NTH-ORDER CORRELATIONS

total number of unknown variables, the presented set of values describes an optimal set of measurements.

N odd

If N is odd, the previously described approach does not work, since α and α' in Eq. (2.21) are both even or odd. Because of this, different correlation terms will depend on the same power of $\cos\theta$ and it is consequently only possible to solve for their sum. Therefore, we modify the choice of our values of ϕ and θ slightly: we choose $N+2$ settings for ϕ, with $\phi_k = \frac{2\pi k}{N+2}$, $k = \{1, 2, \ldots, N+2\}$, so that $e^{i\phi_k}$ describes all $(N+2)$th roots of unity. For every ϕ_k, we conduct measurements for N different values of θ, with $\theta_j = \frac{j}{N+1}\frac{\pi}{2}$ and $j \in \{1, 2, \ldots, N\}$. For this set of $N(N+2)$ measurements, we proceed as in the case for even N: the sum of all measurements for constant θ yields

$$\frac{1}{N+2} x_{\theta_j} = \sum_{\alpha \in G_\beta} \binom{N}{\frac{\alpha}{2}} \binom{N}{\frac{\alpha}{2}} (\cos\theta_j)^{2N-\alpha} (\sin\theta_j)^\alpha \langle \hat{a}_1^{\dagger N-\frac{\alpha}{2}} \hat{a}_2^{\dagger \frac{\alpha}{2}} \hat{a}_1^{N-\frac{\alpha}{2}} \hat{a}_2^{\frac{\alpha}{2}} \rangle, \qquad (2.22)$$

which is identical to Eq. (2.20). However, we have only N equations to solve for $N+1$ terms, so we must conduct one more measurement [e.g. for $(\theta, \phi) = (0, 0)$] to solve for all unknowns. At this point, the total number of measurements is again $N(N+2) + 1 = (N+1)^2$ and, thus, also optimal. In the following, we need not make more measurements but can directly solve for the remaining unknown variables. In a first step, we multiply all equations by $e^{i\phi_1}$ before the summation and arrive at

$$\frac{e^{i\phi_1}}{N+2} x_{\theta_j} = \sum_{\alpha \in G_\beta} \binom{N}{\frac{\alpha-1}{2}} \binom{N}{\frac{\alpha+1}{2}} (\cos\theta_j)^{2N-\alpha} (\sin\theta_j)^\alpha \langle \hat{a}_1^{\dagger N-\frac{\alpha+1}{2}} \hat{a}_2^{\dagger \frac{\alpha+1}{2}} \hat{a}_1^{N-\frac{\alpha-1}{2}} \hat{a}_2^{\frac{\alpha-1}{2}} \rangle. \qquad (2.23)$$

In contrast to the case for even N, we can arrive at a system of equations similar to Eq. (2.22) with only N different terms, which we can solve immediately. Multiplying all equations with $e^{i\phi_k}$ with $2 \leq k \leq (N+1)/2$ gives all other necessary equations in a form equivalent to Eq. (2.21):

$$\frac{e^{i\phi_k}}{N+2} x_{\theta_j} = \sum_{\alpha \in G_\beta} \binom{N}{\frac{\alpha+\beta}{2}} \binom{N}{\frac{\alpha-\beta}{2}} (\cos\theta_j)^{2N-\alpha} (\sin\theta_j)^\alpha \langle \hat{a}_1^{\dagger N-\frac{\alpha-\beta}{2}} \hat{a}_2^{\dagger \frac{\alpha-\beta}{2}} \hat{a}_1^{N-\frac{\alpha+\beta}{2}} \hat{a}_2^{\frac{\alpha+\beta}{2}} \rangle$$
$$+ \sum_{\alpha' \in G_{\beta'}} \binom{N}{\frac{\alpha'+\beta'}{2}} \binom{N}{\frac{\alpha'-\beta'}{2}} (\cos\theta_j)^{2N-\alpha'} (\sin\theta_j)^{\alpha'} \langle \hat{a}_1^{\dagger N-\frac{\alpha'-\beta'}{2}} \hat{a}_2^{\dagger \frac{\alpha'-\beta'}{2}} \hat{a}_1^{N-\frac{\alpha'+\beta'}{2}} \hat{a}_2^{\frac{\alpha'+\beta'}{2}} \rangle, \qquad (2.24)$$

again with $\beta = k$, but $\beta' = N - k + 2$. Since N is odd, all terms now depend on a different power of N, so that the system of linear equations again corresponds to a solvable $(N+1) \times (N+1)$ matrix, making it possible to determine all remaining correlations. For $N = 1$ (i.e., simply a polarization measurement), this leads to the choice of measuring the averages

$$\begin{gathered}
\langle \hat{a}_1^\dagger \hat{a}_1 \rangle, \\
\langle \hat{a}_1^\dagger \hat{a}_1 \rangle + \langle \hat{a}_2^\dagger \hat{a}_2 \rangle + e^{i\frac{2\pi}{3}} \langle \hat{a}_1^\dagger \hat{a}_2 \rangle + e^{-i\frac{2\pi}{3}} \langle \hat{a}_2^\dagger \hat{a}_1 \rangle, \\
\langle \hat{a}_1^\dagger \hat{a}_1 \rangle + \langle \hat{a}_2^\dagger \hat{a}_2 \rangle + e^{i\frac{4\pi}{3}} \langle \hat{a}_1^\dagger \hat{a}_2 \rangle + e^{-i\frac{4\pi}{3}} \langle \hat{a}_2^\dagger \hat{a}_1 \rangle, \text{ and} \\
\langle \hat{a}_1^\dagger \hat{a}_1 \rangle + \langle \hat{a}_2^\dagger \hat{a}_2 \rangle + \langle \hat{a}_1^\dagger \hat{a}_2 \rangle + \langle \hat{a}_2^\dagger \hat{a}_1 \rangle,
\end{gathered} \quad (2.25)$$

which is different from what is discussed in standard textbooks [42], but equally optimal.

2.5.3 Example

In this section, we discuss the simplest nontrivial example, namely measuring all second-order correlations (case $N = 2$). For this task, we start by using the notation of Simon and Mukunda [37] and write the unitary transformation of Eq. (2.7) in terms of three Euler angles ξ, η, ζ:

$$U(\xi, \eta, \zeta) = \exp\left(-i\frac{\xi\sigma_2}{2}\right) \exp\left(i\frac{\eta\sigma_3}{2}\right) \exp\left(-i\frac{\zeta\sigma_2}{2}\right), \quad (2.26)$$

where σ_2 and σ_3 are Pauli matrices. Simon and Mukunda show that the relation of the Euler angles to the actual angles of the three birefringent plates is then given by [37]

$$\alpha_{\text{QP}_1} = \frac{\xi}{2} + \frac{\pi}{4}, \quad (2.27a)$$

$$\alpha_{\text{QP}_2} = \frac{\xi + \eta}{2} + \frac{\pi}{4}, \quad \text{and} \quad (2.27b)$$

$$\alpha_{\text{HP}} = \frac{\xi + \eta - \zeta}{4} - \frac{\pi}{4}, \quad (2.27c)$$

with the Euler angles ξ, η, and ζ as a function of the abstract angles θ and ϕ given by

$$\cos\frac{\eta}{2} = \sqrt{a^2 + c^2},$$

$$\exp\left(i\frac{\xi + \zeta}{2}\right) = \frac{c - ia}{\sqrt{a^2 + c^2}}, \quad \text{and}$$

$$\exp\left(i\frac{\xi - \zeta}{2}\right) = \frac{ib}{|b|},$$

where $a = \text{Re}(e^{i\phi})\sin\theta$, $b = \text{Im}(e^{i\phi})\sin\theta$, and $c = \cos\theta$. In the case that $a = c = 0$, which occurs for $\theta, \phi = \pm\frac{\pi}{2}$, the corresponding Euler angles may be chosen as $\eta = \xi = 0$ and $\zeta = 2\phi$, while in case that $b = 0$, which occurs for $\phi = 0, \pi$, the corresponding Euler angles may be chosen as $\eta = \xi = 0$ and $\zeta = -2\theta$ ($\zeta = 2\theta$) for $\phi = 0$ ($\phi = \pi$).

(θ, ϕ)	Euler angles (ξ, η, ζ)	$(\alpha_{\text{QP}_1}, \alpha_{\text{QP}_2}, \alpha_{\text{HP}})$
$(\frac{1}{8}\pi, \frac{2}{3}\pi)$	$(1.775, 0.676, 4.197)$	$(1.673, 2.011, 0.169)$
$(\frac{1}{4}\pi, \frac{2}{3}\pi)$	$(2.034, 1.318, 5.176)$	$(1.802, 2.461, 0.329)$
$(\frac{3}{8}\pi, \frac{2}{3}\pi)$	$(-3.833, 1.855, -0.692)$	$(2.010, 2.938, 0.464)$
$(\frac{1}{8}\pi, \frac{4}{3}\pi)$	$(4.917, 0.676, 1.775)$	$(0.102, 0.440, 1.740)$
$(\frac{1}{4}\pi, \frac{4}{3}\pi)$	$(-1.107, 1.318, -4.249)$	$(0.232, 0.891, 1.900)$
$(\frac{3}{8}\pi, \frac{4}{3}\pi)$	$(5.591, 1.855, 2.450)$	$(0.439, 1.367, 2.034)$
$(\frac{1}{8}\pi, 0)$	$(0, 0, -\frac{1}{4}\pi)$	$(\frac{1}{4}\pi, \frac{1}{4}\pi, \frac{13}{16}\pi)$
$(\frac{1}{4}\pi, 0)$	$(0, 0, -\frac{1}{2}\pi)$	$(\frac{1}{4}\pi, \frac{1}{4}\pi, \frac{7}{8}\pi)$
$(\frac{3}{8}\pi, 0)$	$(0, 0, -\frac{3}{4}\pi)$	$(\frac{1}{4}\pi, \frac{1}{4}\pi, \frac{15}{16}\pi)$

Table 2.1: All nine values of the Euler angles ξ, η, and ζ and the angles of the three wave plates in dependence of the parameters θ and ϕ for measuring second-order correlations.

With this translation of two abstract parameters into experimental quantities, it is now straightforward to calculate the settings for our wave plates for any arbitrary measurement. For example, if we wish to measure the nine variances of the Stokes parameters (Eq. (2.13)), the recipe from the previous section tells us to measure the second-order intensity after application of the nine unitary transformations that arise from all possible combinations of $\theta = \frac{1}{8}\pi, \frac{1}{4}\pi, \frac{3}{8}\pi$ and $\phi = \frac{2}{3}\pi, \frac{4}{3}\pi, 0$. According to Eqs. (2.27), every pair (θ, ϕ) corresponds to a certain triple of Euler angles (ξ, η, ζ) and this in turn to a certain triple of angles for the wave plates (QP$_1$,QP$_2$,HP), all of which are given in Table 2.1.

From this set of measurements of the second-order intensity moment, it is possible to calculate all variances of the Stokes parameters. In the case of a two-photon Fock state, this corresponds to all density matrix elements (i.e., to a full state tomography). However, our method also serves for the determination of higher-order correlations of other (classical or nonclassical) states. For example, recently the covariance matrix of a Gaussian output state of an optical parametric oscillator has been measured [23]. If one would want to verify the Gaussian property of this state, the measurement of higher-order correlations like the ones discussed in the present work is required.

2.6 Conclusion

In conclusion, it was shown that a very simple experimental setup consisting of two quarter-wave plates, one half-wave plate, a polarizing beam splitter, and a measurement of higher-order intensity moments allows for an optimal measurement of arbitrary-order correlations between two orthogonally polarized modes in a single light beam. Explicit formulas are given for the settings of the three involved wave plates. With these settings, the measurements allow the correlations to be obtained by a solvable system of linear equations. The concept has been exemplified for the case $N = 2$, whereby, in the case of a Fock state, the capability of the method to perform a full state tomography has been demonstrated.

In contrast to a scheme to measure Nth-order correlations proposed by Shchukin and Vogel, we only need to measure Nth-order intensities, instead of $2N$th-order intensities [45]. As a drawback, we cannot measure phase-sensitive moments like $\langle \hat{a}_1 \hat{a}_2 \rangle$. However, the scheme could also be extended to include the measurement of phase-sensitive moments; in this case, one would need to add a local oscillator before the detector in Fig. 2.1. In a recent paper [23], the measurement of such phase-sensitive expectation values was reported for a Gaussian state. For the verification of the Gaussian property of such a state, the measurement of the Nth-order intensity moments like the ones introduced in this chapter is required.

The scheme presented here is also not limited to photons of linear polarization: $\hat{a}_1$ and $\hat{a}_2$ may just as well correspond to any other pairwise orthogonal photon polarization modes. In fact, the general idea is applicable to any kind of bosonic multiqubit state where the equivalents to the needed devices exist: a universal SU(2) gadget, a filter which transmits only one of the two qubit states, and a detector capable of performing a correlation measurement on the incident qubits. For example, one possible application could be the characterization of Laguerre-Gaussian beams with photons distributed among two different LG_{nm} modes [35, 36]. In this case, Agarwal discussed the SU(2) structure of their Poincaré sphere [46]; the equivalent of a polarizing beam splitter can be implemented with holograms [47], and Ref. [48] points at the possibility of constructing an SU(2) gadget consisting of astigmatic lenses.

Chapter 3

Projective Measurements: Quantum State Engineering

3.1 Introduction

In the previous chapter, we investigated how to measure correlations of the electromagnetic field in a single light beam. In the following, we will again discuss correlation measurements of the electromagnetic field. However, in contrast to the previous chapter, we do not restrict the considerations to a single spatial mode anymore but instead discuss the measurement of correlations of an electromagnetic field with contributions from various modes and sources. Nevertheless, while broadening our perspective with respect to the number of spatial modes, we will narrow the view with respect to the sources of the electromagnetic field, which will consist only of a small number of single photon sources.

In contrast to the previous chapter, these correlation measurements are not investigated as an end in themselves, but serve as a tool namely to perform projective measurements in order to generate families of quantum states. It is well known that by detecting photons one can create entanglement among the photons via postselection (cf. e.g. [49]). As first proposed by Bose *et al.* and Cabrillo *et al.*, this technique can also serve to create entanglement between the emitters [8,9]. The generation of entangled states is of interest because they play a key role in the investigations of fundamental aspects of quantum mechanics [50,51] and they are widely used as a basic resource for different tasks in quantum information processing [52], e.g., for applications in quantum cryptography [53,54], quantum teleportation [55], or quantum computing [3,56].

Entangled photons are easily transported over large distances which is of interest whenever entanglement is to be distributed. In addition, parametric down conversion provides for very bright sources of single photons so that entangled photons can be created at a very fast rate, enabling to create entanglement very quickly [24].

On the other hand, if projective measurements are used to entangle the emitters, these emitters may be separated by arbitrary distances, as there is no need for a direct interaction between them. This should be contrasted with other schemes entangling massive particles, which require some kind of interaction, be it Coulomb-like [57–60] or mediated by photons [61–65]. Furthermore, at variance with photon entanglement, usually achieved by parametric down conversion [7, 24–27, 29, 49, 66–76], the entanglement of electronic ground states of atoms can be preserved over long periods of time [59, 60]. By now, this technique has been proposed for the creation of a large number of entangled states [77–82] and Moehring et al. recently realized it for the first experimentally with two ^{171}Yb$^+$ ions as photon emitters [83]. The technique may find further applications, in particular in quantum communication and quantum computation, where long-lived entanglement plays a crucial role.

In the present work, we will mainly concentrate on the entanglement via projective measurements of photons of the emitters themselves, i.e., the single photon sources. We will show that it is possible to create not only maximally entangled states among the emitters but also states entangled to an arbitrary degree. Furthermore, we will describe setups which are able to create two families of states, the family of total angular momentum eigenstates and and a family of clusters states. Afterwards, we will introduce a comprehensive approach which investigates the possibilities and limitations of projective measurement-based quantum state generation. Finally, we will demonstrate how projective measuerement-based quantum state generation in the photon emitters is related to the projective measurement-based quantum state generation in the photons themselves. By introducing this connection, it becomes possible to transfer all techniques that have been developed straightforwardly to the generation of the same families of states among the emitted photons, rather than among the emitters.

3.2 Tool Box

3.2.1 The Λ-level System

The basic idea of this chapter is to demonstrate the entanglement of non-interacting localized emitters by measuring the photons they emit. Conveniently, one might use ions as emitters, because they can easily be trapped and localized. In addition, all photons emitted by different ions of the same atomic species are indistinguishable, a feature which is necessary for our considerations. Therefore, in the following, we will consider trapped atoms, keeping in mind that only the energy level structure is relevant and all considerations will remain valid for other light sources wih a suitable internal structure such as neutral atoms, quantum dots, molecules, NV centers in diamond, etc..

Throughout this chapter, we will consider systems which have three separate energy eigenstates $|e\rangle$, $|+\rangle$, and $|-\rangle$. $|+\rangle$ and $|-\rangle$ are supposed to be degenerate and stable in

3.2. TOOL BOX

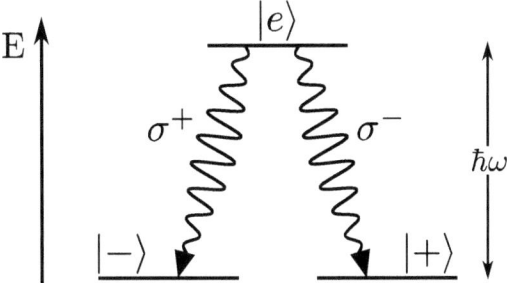

Figure 3.1: Schematic drawing of a system with a Λ-level energy structure. The two ground states are degenerate and seperated by $\hbar\omega$ from the excited state $|e\rangle$.

the sense that there are no lower energy levels into which $|\pm\rangle$ decay on a timescale relevant for the discussion. These two states will serve to encode a qubit. The excited state $|e\rangle$ has an energy $\hbar\omega$ where the frequency ω is in or close to the optical domain. Because of the similarity of the graphical representation of the level structure (cf. Fig. 3.1) to a capital greek letter, the level structure is said to be arranged in a Λ configuration, or in short, that it is a Λ-level system.[1] In addition, for our scheme we require that the polarization of the photons emitted on the transition from the excited state to each of the ground states is orthogonal and that both transitions occur with equal probability. Physically, such a system may be realized by using the Zeeman sublevels of an energy eigenstate of an atom. In such a case, a photon with polarization σ^+ is emitted if the atom decays into its $|-\rangle$ state and a σ^--polarized photon is emitted if the atom decays into the $|+\rangle$ state.

3.2.2 Atom-Photon Entanglement

Following Agarwal [85] who derived the electromagnetic field of a two-level atom, the positive frequency part of the electric field at position **r** generated by a Λ-level atom in the far field and rotating wave approximation is found to be

$$\hat{\mathbf{E}}^+(\mathbf{r},t) \propto -k_0^2 \frac{\mathbf{R} \times (\mathbf{R} \times \mathbf{d})}{R} e^{ik_0 \mathbf{r} \cdot \mathbf{R}} (\boldsymbol{\varepsilon}_{\sigma^-}|+\rangle\langle e| + \boldsymbol{\varepsilon}_{\sigma^+}|-\rangle\langle e|), \qquad (3.1)$$

where **R** denotes the position of the atom, k_0 the wave number of its resonance frequency, $\boldsymbol{\varepsilon}_{\sigma^\pm}$ is again the unit vector for the $\sigma^\pm$-polarization, and **d** is the dipole moment of the atom. Not surprisingly, the state of the field is determined by the state of the atom. In other words: in order to emit a photon, the atom has to be transferred

---
[1]Feng et al. first discussed the particular usefulness of this type of system in the context of measurement-based quantum state generation [84].

into the excited state. Subsequently, it decays either into its $|+\rangle$ state and emits a σ^--polarized photon or into its $|-\rangle$ state and emits a σ^+-polarized photon. This decay can be treated with the Weisskopf-Wigner method for spontaneous emission, in which the atom-photon system is in a time-dependent superposition with the atom either having emitted a photon or not:

$$|\Psi\rangle(t) = a(t)|e\rangle_a|0\rangle_p + b(t)\left(|+\rangle_a|\sigma^-\rangle_p + |-\rangle_a|\sigma^+\rangle_p\right), \quad (3.2)$$

where the first ket denotes the state of the atom and the second ket that of the photon. In the following, we will assume that the atom has decayed from the excited state. Experimentally, this could be assured by either waiting long enough such that the factor $a(t)$ which describes an exponential decay, is practically zero or by postselectively including only those events in which a photon is measured. With these assumptions, Eq. (3.2) simplifies to

$$|\Psi\rangle = \frac{1}{\sqrt{2}}\left(|+\rangle_a|\sigma^-\rangle_p + |-\rangle_a|\sigma^+\rangle_p\right). \quad (3.3)$$

This state can easily be shown to be maximally entangled, e.g., by calculating the concurrence (cf. Sec. 3.3.2) and this entanglement has also been verified experimentally [86, 87]. It is this non-local correlation between the atom and the photon that will allow us to manipulate the state of the atom by acting on the photon.

3.2.3 Projective Measurements of Entangled Systems

In quantum mechanics, a measurement is described by measuring some observable X specified by a linear operator $\hat{X}$. The only values that can be found as a result of the measurement are the eigenvalues X_i of this operator. Assuming that they are non-degenerate, each eigenvalue corresponds to exactly one eigenstate of said observable, into which the system is projected upon the measurement. This is a stochastic procedure of preparing certain states, with a simple example already mentioned in the introduction. If we start with an entangled state, then it is even possible to project one subsystem, i.e., prepare it in a certain state, by measuring another without the need for any interaction. A simple example can be constructed from Eq. (3.3): By detecting the photon with a certain polarization $|p\rangle = \alpha^*|\sigma^-\rangle + \beta^*|\sigma^+\rangle$, one projects the state of the photon and of the atom while the atom is not subjected to any interaction. Thus, the projected state has the following form:

$$|\Psi^{(f)}\rangle = (\mathbb{1}_a \otimes |p\rangle_p\langle p|)|\Psi\rangle = (\alpha|+\rangle_a + \beta|-\rangle_a) \otimes |p\rangle_p \quad (3.4)$$

Obviously, the atomic qubit can be prepared in any desired state without having to interact with the measurement device which detects (and projects) only the state of

3.2. TOOL BOX

the photon. In this way, it is possible to remotely engineer a certain quantum state, of course with the drawback that the generation is only probabilistic. One trades this drawback for the huge advantage that there is no need to control any interaction processes.

3.2.4 Measuring Photons from more than one Atom

Switching from a setup where only a single atom scatters photons to a setup with two atoms gives yet more possibilities. Suppose a detector D_1 is placed at some position $\mathbf{r}_1$ and the polarization filter in front of it allows only σ^--polarized photons to pass. If this detector now registers a photon, we do not know by which of the atoms it was emitted and, consequently, have to sum over both possibilities. In addition, when propagating from each of the atoms towards the detector, then according to Eq. (3.1), the photon accumulates a certain optical phase, φ_{11} and φ_{21}, which also has to be taken into account. Behind of the polarization filter the state of the atom-photon system therefore is of the form

$$|\Psi^{(1)}\rangle = \left(e^{i\varphi_{11}}|+e\rangle_a + e^{i\varphi_{21}}|e+\rangle_a\right) \otimes |\sigma^-\rangle_p. \quad (3.5)$$

This is a state, in which the two atoms are entangled with each other without ever having interacted. However, as one atom is still excited, the state will decay quickly. Now suppose a second detector D_2 at $\mathbf{r}_2$ registers a σ^+-polarized photon. In that case, $|\Psi^{(1)}\rangle$ is projected again, this time with the optical phases φ_{12} and φ_{22}. The resulting state can be written as

$$|\Psi^{(2)}\rangle = \left(e^{i(\varphi_{11}+\varphi_{22})}|+-\rangle_a + e^{i(\varphi_{21}+\varphi_{12})}|-+\rangle_a\right) \otimes |\sigma^-\sigma^+\rangle_p. \quad (3.6)$$

As the atoms are in their ground state, the lifetime of this entangled state is in principle unlimited. Furthermore, the entanglement has been generated remotely, i.e., without any interaction with the atoms themselves. Note that in order to have well defined phases φ_{jk}, it is important that the geometry of the detector is chosen such that the optical path length from any atom to any point on its surface does not vary appreciably.

This scheme can be generalized to an arbitrary number N of atoms and arbitrary orientations of polarization filters in front of the detectors. The experimental approach is then described as follows: The N atoms are concurrently excited by a laser pulse and the photons they scatter are detected by N detectors positioned at $\mathbf{R}_i$. The setup is inherently probabilistic, since one can not be sure that a photon is detected at any, let alone all detectors. Therefore, the excitation has to be repeated until a successful measurement event occurs, i.e., until N photons are registered, exactly one at each detector. In each measurement cycle, the trapped ions are transferred to the excited

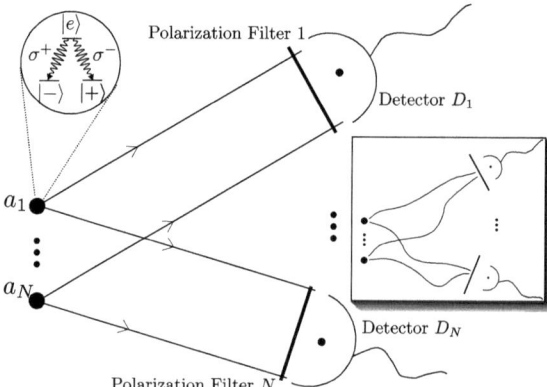

Figure 3.2: Proposed setup to entangle remote emitters with an internal Λ-level scheme: N photons emitted by N single photon emitters (preferably atoms) are detected with N single photon detectors equipped with polarization filters, either in the far field of the emitters or with the photons guided by optical fibers to the detectors (inset).

state by a laser π pulse. After that the detectors wait for a time much longer than the lifetime of the excited state for the arrival of a photon. If any detector does not register a photon within this time frame, the next cycle starts.

The scattered photons either propagate through free space towards the detectors or fibers are used to guide them in order to achieve more control and influence different degrees of freedom (cf. Fig. 3.2) [83, 87, 88]. Note that in case that fibers are used the drawing is oversimplified; two fibers ending at one detector cannot simply be connected to that detector independendently, but their signals have to overlap, e.g. on a beam splitter, in order to define a phase between their signals.

In this chapter, we will extensively use this projective technique to generate entanglement between two or more atoms and to engineer a large variety of quantum states.

3.2.5 The Detection Operator

In the last two sections, we introduced the technique of projective measurements of photons to generate a certain state among the atoms. In the initial state the atoms are completely excited while the field is in its vacuum state $|0\rangle_p$. In a unitary evolution process, the field modes are subsequently populated and the excited state of the atoms is depopulated (cf. Eq. (3.2)). As photons are usually measured destructively, a successful measurement cycle in which all scattered photons are registered projects the field back

3.2. TOOL BOX

into the vacuum state[2]. In order to simplify the notation, we therefore introduce a detection operator $\hat{D}_k$ which is applied to the state of the atoms if the kth detector registers a photon and which describes the projection of the atoms directly and without reference to the state of the field. Omitting proportionality factors, it is given by

$$\hat{D}_k = \hat{D}_k(\mathbf{r}_k) = \sum_{j=1}^{N} e^{i\varphi_{jk}} \left(\alpha_j |+\rangle_{a_j}\langle e| + \beta_j |-\rangle_{a_j}\langle e| \right) \otimes \mathbb{1}_{\neg a_j}, \qquad (3.7)$$

where the sum runs over all N atoms and the prefactors α_j and β_j are determined by the contribution of the respective atomic transition to the field amplitude at the detector position behind the polarization filter; with no other optical elements present they are thus completely determined by the orientation of the polarization filter $\boldsymbol{\varepsilon} = \alpha\boldsymbol{\varepsilon}_{\sigma^-} + \beta\boldsymbol{\varepsilon}_{\sigma^+}$. The symbol $\mathbb{1}_{\neg a_j}$ denotes the unit operator on all atoms except for atom a_j; in order not to overcrowd the equations, it will from now on be always implicitly assumed whenever the explicit notation of an operator does not have the same dimension as the state space to which the operator is applied. φ_{jk} is the phase that the photon accumulates from source j to detector k. With such a detection operator, the state of the atoms can be described by a simple von Neumann projection introduced in Eq. (1.2):

$$|\phi^{(f)}\rangle = \frac{\hat{D}_k|\phi^{(i)}\rangle}{\sqrt{\langle\phi^{(i)}|\hat{D}_k^\dagger \hat{D}_k|\phi^{(i)}\rangle}}. \qquad (3.8)$$

Since all state vectors refer to the state of the atoms, the subscript from the previous sections has been dropped.

With these definitions, the probability that a detector actually clicks follows directly from Eq. (1.1) as

$$P(k) = A_k \langle\phi^{(i)}|\hat{D}_k^\dagger \hat{D}_k|\phi^{(i)}\rangle, \qquad (3.9)$$

with A_k a proportionality factor which is determined by the specific experimental setup and in which enter factors like the quantum efficiency of the detector and the numerical aperture of the photon collection system. It could be absorbed in the detection operator by defining $\hat{D}_k' = \sqrt{A_k}\hat{D}_k$, but one can immediately see from Eq. (3.8) that any rescaling of $\hat{D}_k$ is irrelevant for the state that is created. Therefore, A_k will be omitted in the following.

With these tools at hand, we are now ready to discuss the possibilities of projective measurements for the purpose of quantum state generation.

---

[2] Even if the photons are measured non-destructively, their state will be completely determined and separable from the state of the atoms which we are interested in.

3.3 Heralded Entanglement of Arbitrary Degree in Remote Qubits [11]

3.3.1 The Idea

In this section, we present a proposal involving a simple scheme to operationally tune the amount of long-lived entanglement present in two remote atomic qubits. We will demonstrate that our scheme allows to create heralded entangled states, where the degree of entanglement between the atomic qubits can be tuned at will. The exact value is determined by adjusting two easily accessible experimental parameters, namely the position of two photodetectors and the relative orientation of two polarizers.

In the following, we study the setup described in Sec. 3.2.4 for the case of $N = 2$ atoms. In a free space configuration, this installation can be considered as a Young-type interferometer realized by the two localized atoms [89] with an internal Λ-level structure (see Fig. 3.3). The atoms, representing the double-slit of the interferometer, are excited by a laser pulse and subsequently scatter photons in their deexcitation process. These photons are registered in the far field with polarization sensitive photon detectors where the far-field detection is chosen for the practical reasons.[3]

According to Eq. (3.7) and Ref. [90], the detection of a photon scattered off two Λ-level atoms a_1 and a_2, with a detector D_i at position $\mathbf{r}_i$ behind a polarization filter aligned along $\boldsymbol{\varepsilon}_i$, is described by the projection operator

$$\hat{D}_i = \hat{D}_i(\mathbf{r}_i) = \sum_{m=\pm} (\boldsymbol{\varepsilon}_i \cdot \mathbf{d}_{me}) \left(|m\rangle_{a_1}\langle e| + e^{-i\varphi(\mathbf{r}_i)} |m\rangle_{a_2}\langle e| \right), \qquad (3.10)$$

where the sum runs over the two ground states $|\pm\rangle$. Here, $\mathbf{d}_{\pm e}$ is the dipole moment of the transition $|e\rangle \to |\pm\rangle$ and, according to Eq. (3.1), the phase difference $\varphi(\mathbf{r}_i)$ in the far field approximation is given by

$$\varphi(\mathbf{r}_i) = k(\mathbf{R}_{a_2} - \mathbf{R}_{a_1}) \cdot \mathbf{e}(\mathbf{r}_i), \qquad (3.11)$$

where k is the wavenumber of the detected photon, $\mathbf{R}_{a_1,a_2}$ is the position of the respective atom, and $\mathbf{e}(\mathbf{r}_i)$ is the unit vector pointing from the atoms towards the detector D_i at $\mathbf{r}_i$. Hereby, the far-field detection ensures that $\mathbf{e}(\mathbf{r}_i)$ is identical for both atoms. The same measurement can be accomplished by using optical fibers guiding the photons

---

[3] In the near field several issues complicate the experimental implementation: for one, basic geometric considerations show that the variations of the phase $\varphi(\mathbf{r})$ introduced below in Eq. (3.11) varies on the order of the wavelength of the photons. However, in order to have a well-defined phase, it is necessary that the phase does not vary appreciably over the surface of the detector. At the same time, the detector must be massive enough that the momentum transfer from the photon is negligible so that the experimenter is not able to gain any path information from the momentum transfer (compare Einstein's recoiling slit (Ch. 4)). Thirdly, the field amplitudes from both emitters differs in the near field typically by a non-negligible amount and would also have to be accounted for.

3.3. HERALDED ENTANGLEMENT IN REMOTE QUBITS

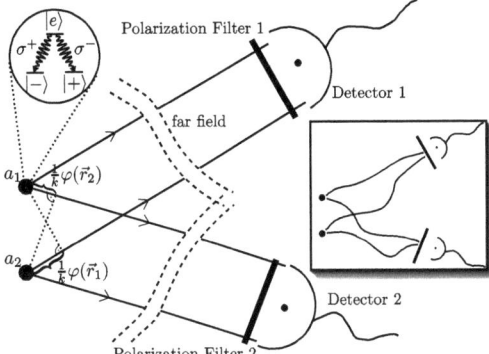

Figure 3.3: Scheme of two atoms with internal Λ-level structures using two detectors in the far field with polarization filters in front to register the photons emitted by the atoms. The inset shows the same configuration using optical fibers.

from the atoms to the detectors [83, 87, 91]. In this case, the phase difference $\varphi(\mathbf{r}_i)$ is given by

$$\varphi(\mathbf{r}_1) = k\left(w_{a_2}(\mathbf{r}_1) - w_{a_1}(\mathbf{r}_1)\right), \quad (3.12)$$

where $w_{a_1,a_2}(\mathbf{r}_i)$ is the optical path length from the respective atom to the detector D_i at $\mathbf{r}_i$ via the corresponding optical fiber (cf. Fig. 3.3). This more general expression for $\varphi(\mathbf{r}_1)$ describes all cases in which the far field condition is not met for any reason, thus including also situations in which the atoms are separated by arbitrary distances, i.e., for truly remote particles.

By applying the operator $\hat{D}_1$ and $\hat{D}_2$ to the initial double excited state of the two atoms, $|\psi_\Lambda^{(i)}\rangle = |ee\rangle$, we find the normalized atomic state after the detection of two photons to be

$$\begin{aligned}|\psi_\Lambda^{(f)}\rangle &= \frac{\hat{D}_1\hat{D}_2|\psi_\Lambda^{(i)}\rangle}{\sqrt{\langle\psi_\Lambda^{(i)}|\hat{D}_2^\dagger\hat{D}_1^\dagger\hat{D}_1\hat{D}_2|\psi_\Lambda^{(i)}\rangle}} = \\ &\frac{1}{\sqrt{\zeta}}\Big[\left(1+e^{-i\varphi_{21}}\right)\left(\varepsilon_{2-}\varepsilon_{1-}|++\rangle + \varepsilon_{2+}\varepsilon_{1+}|--\rangle\right) \\ &+ \left(\varepsilon_{2+}\varepsilon_{1-} + e^{-i\varphi_{21}}\varepsilon_{2-}\varepsilon_{1+}\right)|-+\rangle \\ &+ \left(e^{-i\varphi_{21}}\varepsilon_{2+}\varepsilon_{1-} + \varepsilon_{2-}\varepsilon_{1+}\right)|+-\rangle\Big]. \quad (3.13)\end{aligned}$$

Here, the abbreviation $\varepsilon_{i\pm} = \boldsymbol{\varepsilon}_i \cdot \mathbf{d}_{\mp e}$ is used, where without loss of generality we assume $|\varepsilon_{i+}|^2 + |\varepsilon_{i-}|^2 = 1$, φ_{21} is given by the phase difference

$$\varphi_{21} = \varphi(\mathbf{r}_2) - \varphi(\mathbf{r}_1), \quad (3.14)$$

depending on the two detector positions $\mathbf{r}_1$ and $\mathbf{r}_2$, and $\zeta = 2(1 + |\boldsymbol{\varepsilon}_2 \cdot \boldsymbol{\varepsilon}_1^*|^2 \cos\varphi_{21})$ is a normalization factor.

3.3.2 Quantifying the Entanglement: A Malus' Law for the Concurrence

Out of a large number of entanglement measures that have been introduced, only the logarithmic negativity and the concurrence are easily computable for an arbitrary state of two qubits [92]. For the present purpose, the concurrence is best suited since it has a particularly simple form for pure states. According to Wootters [93], if one defines the spin-flipped state $\tilde{\rho}$ for an arbitrary state ρ of two qubits as

$$\tilde{\rho} = (\sigma_{y,a_1} \otimes \sigma_{y,a_2}) \rho^* (\sigma_{y,a_1} \otimes \sigma_{y,a_2}), \qquad (3.15)$$

where $\sigma_{y,X}$ is the usual σ_y Pauli matrix of qubit X ($X = a_1, a_2$), then the concurrence is given as

$$\mathcal{C}(\rho) = \max\{0, \lambda_1 - \lambda_2 - \lambda_3 - \lambda_4\} \qquad (3.16)$$

with $\lambda_{1,2,3,4}$ the four square roots of the (real and positive) eigenvalues of $\rho\tilde{\rho}$ in descending order. For an arbitrary pure state of two qubits $|\psi\rangle = a|++\rangle + b|+-\rangle + c|-+\rangle + d|--\rangle$ this simplifies to [93]

$$\mathcal{C}(|\psi\rangle) = |\langle \tilde{\psi}|\psi\rangle| = 2|ad - bc|. \qquad (3.17)$$

Using this equation, the concurrence of the pure state from Eq. (3.13) can be explicitly calculated. One obtains

$$\mathcal{C}(\varphi_{21}, \mathcal{V}_{12}) = \frac{|\varepsilon_{2+}\varepsilon_{1-} - \varepsilon_{2-}\varepsilon_{1+}|^2}{1 + |\boldsymbol{\varepsilon}_2 \cdot \boldsymbol{\varepsilon}_1^*|^2 \cos\varphi_{21}} = \frac{1 - \mathcal{V}_{12}}{1 + \mathcal{V}_{12} \cos\varphi_{21}}, \qquad (3.18)$$

where the parameter $\mathcal{V}_{12}$ is given by

$$\mathcal{V}_{12} = |\boldsymbol{\varepsilon}_2 \cdot \boldsymbol{\varepsilon}_1^*|^2. \qquad (3.19)$$

According to Eq. (3.18), the long-lived entanglement generated between the ground states of the two Λ-level atoms only depends on the relative phase φ_{21} and on the relative orientation of the two polarization filters $\mathcal{V}_{12}$ (see Fig. 3.4).[4] In order to obtain a certain amount of entanglement between the two atoms, both parameters have to be tuned to suitable values and the excitation of the atoms has to be repeated until both

---
[4]Note that Eq. (3.18) only holds if $\mathcal{V}_{12}$ and $\cos\varphi_{21}$ are not simultaneously equal to 1 and -1, respectively, since in this case no two-photon signal is detected (cf. Eq. (3.23)) so that the projection described in Eq. (3.13) does not occur.

3.3. HERALDED ENTANGLEMENT IN REMOTE QUBITS

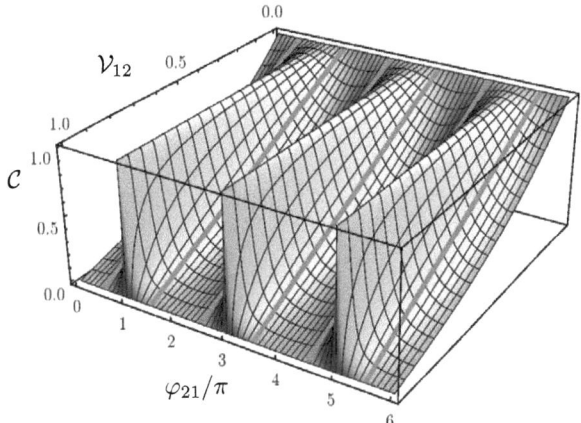

Figure 3.4: The concurrence as a function of the relative phase φ_{21} (scaled in multiples of π) and the parameter $\mathcal{V}_{12}$. The thick orange lines mark constant $\varphi_{21} = (n+1/2)\pi$, where the dependence of $\mathcal{C}$ on $\mathcal{V}_{12}$ becomes linear.

detectors have registered a photon. By postselection we then know that the atomic pair contains exactly the desired amount of entanglement as described by Eq. (3.18). Taking a look at the extremal values of $\mathcal{C}$ with respect to φ_{21}, we find

$$\left.\begin{aligned}\mathcal{C}_{\min} = \mathcal{C}_{\max} &= 0 \quad \text{for all} \quad \cos\varphi_{21}, \quad \text{if } \mathcal{V}_{12} = 1 \text{ and} \\ \mathcal{C}_{\min} &= \tfrac{1-\mathcal{V}_{12}}{1+\mathcal{V}_{12}} \quad \text{for} \quad \cos\varphi_{21} = 1, \\ \mathcal{C}_{\max} &= 1 \quad \text{for} \quad \cos\varphi_{21} = -1,\end{aligned}\right\} \text{if } \mathcal{V}_{12} \neq 1. \quad (3.20)$$

These expressions show that, depending on the value of $\mathcal{V}_{12}$, *any* amount of concurrence between $\tfrac{1-\mathcal{V}_{12}}{1+\mathcal{V}_{12}}$ and 1 can be achieved. In particular, by choosing φ_{21} to be an odd multiple of π, it is always possible to generate a state with maximal (unit) concurrence, independent of the explicit value of $\mathcal{V}_{12} < 1$, i.e., independent of the relative orientation of the two polarization filters as long as the polarizers do not point in the exact same direction (see Fig. 3.4).

The extrema of the concurrence with respect to $\mathcal{V}_{12}$ are given by

$$\left.\begin{aligned}\mathcal{C}_{\min} = \mathcal{C}_{\max} &= 1 \quad \text{for all } \mathcal{V}_{12} < 1, \quad \text{if } \cos\varphi_{21} = -1 \text{ and} \\ \mathcal{C}_{\min} &= 0 \quad \text{for} \quad \mathcal{V}_{12} = 1, \\ \mathcal{C}_{\max} &= 1 \quad \text{for} \quad \mathcal{V}_{12} = 0,\end{aligned}\right\} \text{if } \cos\varphi_{21} \neq -1. \quad (3.21)$$

Thus, if the phase difference is not fixed to an odd multiple of π, it is always possible to use $\mathcal{V}_{12}$ as a single parameter to tune the concurrence to any desired value. In particular,

40   CHAPTER 3. QUANTUM STATE ENGINEERING

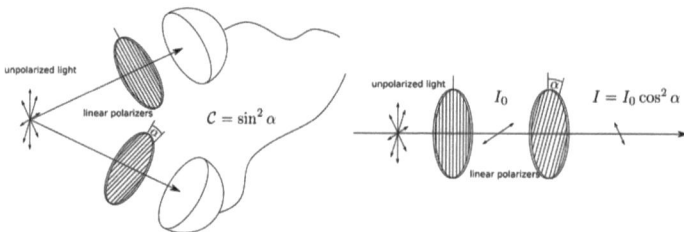

Figure 3.5: For the generation of an arbitrary amount of concurrence (at $\varphi_{21} = 0$) the generated concurrence is proportional to $\sin^2 \alpha$ for *non-collinear* polarizers, i.e., each photon passes only one polarization filter and does not know about the other (left). For the classical Malus' Law, the same photons pass both polarization filters and the probability of passage is proportional to $\cos^2 \alpha$ (right).

by choosing φ_{21} an odd multiple of $\pi/2$, we find in Eq. (3.18) a *linear* relation between the concurrence and the parameter $\mathcal{V}_{12}$. In this case, when linear polarizers are used, by keeping one of them fixed and turning the other by a relative angle α, we are able to implement a fully tunable concurrence $(0 < \mathcal{C} < 1)$

$$\mathcal{C} = 1 - \mathcal{V}_{12} = \sin^2 \alpha, \qquad (3.22)$$

yielding an analog to the Malus' Law [94]. In its classical version, it says that the intensity of a light beam propagating along a single spatial mode and passing consecutively through two linear polarizers in this mode is proportional to the square of the cosine of the relative angle between the two polarizers. Here, we find that the concurrence, a measure characterizing the entanglement of two qubits, behaves in a similar way. Even though each of the two indistinguishable photons propagates along two *different* modes and passes through a polarizer in each individual mode, the degree of entanglement generated between the atoms upon detection of the photons is determined by the relative angle between the two polarizers. This result can be seen as an operational implementation of a tunable measure of entanglement between matter qubits following a simple and intuitive law of classical optics (cf. Fig. 3.5).

We note that the parameter $\mathcal{V}_{12}$ intervenes also in the second-order correlation function $G^{(2)}(\varphi_{21})$, which is proportional to the signal of measuring two photons, one at $\mathbf{r}_1$ and the other at $\mathbf{r}_2$. According to Eq. (3.9), the second-order correlation function reads [90]

$$\begin{aligned} G^{(2)}(\varphi_{21}) &= A_1 A_2 \langle \psi_\Lambda^{(i)} | \hat{D}_2^\dagger \hat{D}_1^\dagger \hat{D}_1 \hat{D}_2 | \psi_\Lambda^{(i)} \rangle \\ &\propto 2 \left(1 + \mathcal{V}_{12} \cos \varphi_{21} \right), \end{aligned} \qquad (3.23)$$

3.3. HERALDED ENTANGLEMENT IN REMOTE QUBITS

In this expression $\mathcal{V}_{12}$ appears as the *visibility* of the $G^{(2)}(\varphi_{21})$-function, revealing the close relationship between quantum interference and entanglement.

3.3.3 Experimental Feasibility

In the following, we will give an estimate of the variation of the concurrence due to experimental uncertainties (see also [81, 82]). The probability to detect a scattered photon is proportional to the solid angle subtended by the detector divided by 4π. By extending the detection area, the detection probability will increase. However, in general this also leads to the accumulated phases being less well defined (cf. Sec. 3.2.4). Thus, there is a trade-off between the count rate of the scattered photons and the error in the concurrence generated in the final state. For estimating errors, we will assume identical rectangular detectors. Let α_D be the azimuthal angular extension of each detector in direction of θ_i, with θ_i the azimuth angle between $\mathbf{e}(\mathbf{r_l})$ and the axis connecting the two atoms, and φ_D the polar angle subtended by each detector perpendicular to the plane of α_D. Then, for small α_D, the probability to detect a randomly emitted photon with one of the two detectors can be approximated to

$$P(\alpha_D, \varphi_D) = \frac{\alpha_D \varphi_D}{4\pi}. \tag{3.24}$$

The count rate R of two-photon detection events is thus given by

$$R = 2r \cdot P(\alpha_D, \varphi_D)^2 \cdot G^{(2)}(\varphi_{21}), \tag{3.25}$$

where r is the repetition rate of the experiment. A factor of 2 appears since either detector, D_1 or D_2, might register the first photon. The count rate is thus maximal if the condition for constructive interference of the second-order correlation function $G^{(2)}(\varphi_{21})$ is fulfilled, i.e., if φ_{21} is an even multiple of π.

The uncertainty in the concurrence $\Delta\mathcal{C}$ is defined by the difference in the concurrence of the density matrix of the actually generated state $\rho^{(g)}$ (calculated according to Eq. (3.16)) and the pure target state $\rho^{(f)} = |\psi_\Lambda^{(f)}\rangle\langle\psi_\Lambda^{(f)}|$:

$$\Delta\mathcal{C} = |\mathcal{C}(\rho^{(g)}) - \mathcal{C}(\rho^{(f)})|. \tag{3.26}$$

$\Delta\mathcal{C}$ is essentially determined by the uncertainty in the orientation of the polarization filters $\Delta\mathcal{V}$ and the uncertainty in the phase $\Delta\varphi_{21}$. With current experimental technology, $\Delta\mathcal{V}$ can be suppressed to the order of 10^{-10} [95]. Thus, $\Delta\mathcal{V}$ is negligible compared to the uncertainty imposed by the phase and will be neglected in the following.

The uncertainty in the phase $\Delta\varphi_{21}$ is governed by two contributions: the solid angle subtended by the detector (determined by α_D and φ_D) and the finite confinement μ of the atoms in the trap. To calculate $\rho^{(g)}$, we thus have to sum up the pure state density

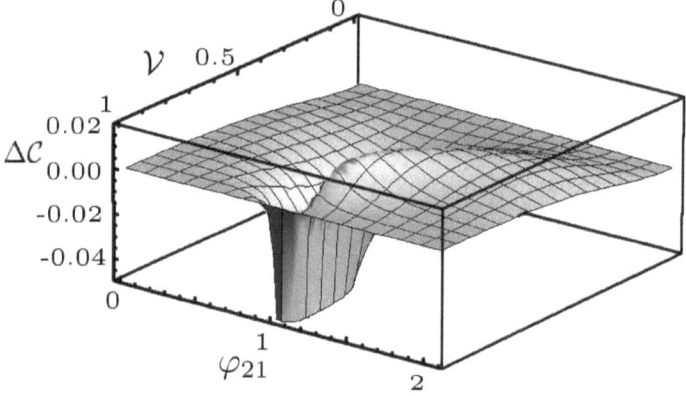

Figure 3.6: Numerical calculation of the error in the concurrence according to Eq. (3.26).

matrices over the whole relevant parameter space:

$$\rho^{(g)} = \int_{\Lambda,\mu} w(\varphi_{21}) \, |\psi_\Lambda^{(f)}\rangle\langle\psi_\Lambda^{(f)}|(\varphi_{21}) \, \mathrm{d}\Lambda \, \mathrm{d}\mu \qquad (3.27)$$

where $w(\varphi_{21})$ is a weight factor determined by the geometry of the setup and normalization. To minimize the deviation from the desired final state, we have to minimize φ_D and α_D, while θ_i should be close to $\frac{\pi}{2}$. However, φ_D and α_D are bound from below by the requirement of an acceptable count rate R. In addition, there is a lower boundary to $\Delta \mathcal{C}$ due to the finite confinement of the atoms.

Numerical calculations show that for realistic experimental parameters, $d = 5\,\mu\mathrm{m}$, $\mu = 10\,\mathrm{nm}$, $\alpha_D = 5\,\mathrm{mrad}$, $\varphi_D = \frac{\pi}{6}$, $\theta \approx \frac{\pi}{2}$, and photons of wavelength $\lambda = 650\,\mathrm{nm}$, this results in $\Delta \mathcal{C}_{\mathrm{max}} < 0.025$ for all $\varphi_{21} \in [-\frac{\pi}{2}, \frac{\pi}{2}]$ and all $\mathcal{V}$ (cf. Fig. 3.6). Within this parameter range, the fidelity of the final state $\rho^{(g)}$ always remains above 95 %, while for a repetition rate r of a few Mhz the count rate amounts to a few events per second. These estimates include a detector efficiency of about 30 % and a dark count rate of up to a few 100 Hz.

The protocol presented here is capable of producing heralded entangled states with a high fidelity. The count rate, on the other hand, is relatively low in the analysed case. Modifications in the setup concerning the detector shape and the number of detectors are possible, as well as the use of fibers or cavities to increase the detection probability of the scattered photons without curtailing the fidelity. These suggested modifications do not change the principal results of this proposal, but they might contribute to a better implementation of the presented basic ideas.

3.3.4 Conclusion

In conclusion, we have shown that with a simple and realistic setup, it is possible to create heralded entanglement of any degree between two remote atoms with a Λ-type level structure. As the atoms are entangled by projective measurements requiring no atomic interaction, the atomic distances in a given experiment are arbitrary. In particular, instead of using a far-field measurement to erase the which-way information of photons, the use of optical fibers could provide a more practical approach to reach similar goals. We expect that our results inspire and stimulate further research in operational and realistic methods for the generation and measure of entanglement in different experimental contexts.

3.4 Generation of Total Angular Momentum Eigenstates in Remote Qubits [10]

3.4.1 Introduction

In the previous section, we have seen that it is possible to generate an arbitrary amount of entanglement between two remote atoms by measuring the photons spontaneously scattered by these two single photon emitters. This scheme is not limited to two atoms. However, for a larger number of atoms no singular measure for the quantification of entanglement exists, since there are types of entanglement inequivalent to each other and therefore incomparable [92,96–98]. In addition, for multipartite systems it is usually the generation of a certain state that is of interest, e.g. cluster states for quantum computation [56]. It was shown earlier in our group that it is possible to generate all symmetric total angular momentum eigenstates (the so called Dicke states after R. H. Dicke [99]) for an arbitrary number of N remote qubits [81] and even all states symmetric under a permutation of the qubits [82] with a setup in which polarization sensitive detectors collect the emitted photons in the far field. In the following, we will show that by using a setup in which fibers guide the photons towards the detectors, it is also possible to generate all non-symmetric total angular momentum eigenstates in N qubits [10]. This shows that with the proposed scheme, all – the symmetric as well as the non-symmetric – total angular momentum eigenstates of an N qubit compound can be produced.

3.4.2 Total Angular Momentum Eigenstates

Total angular momentum eigenstates (TAMEs) are defined as the simultaneous eigenstates of the total angular momentum operator $\hat{\mathbf{J}}^2$ and its z-component $\hat{J}_z$ of a number of physical subsystems. Usually, these operators are used to describe the interaction of the subsystems (e.g. **L-S** coupling) and they have been applied successfully to fields as disparate as solid state, atomic and particle physics. For N spin-1/2 particles, the total angular momentum eigenstates, defined as simultaneous eigenstates of the square of the total spin operator $\hat{\mathbf{S}}^2$ and its z-component $\hat{S}_z$, are commonly denoted by $|S_N;m_N\rangle$, with the corresponding eigenvalues $S_N(S_N+1)\hbar^2$ and $m_N\hbar$ [22,99]. However, since the denomination $|S_N;m_N\rangle$ generally characterizes more than one quantum state [85], we will extend the notation of an N-qubit state by its coupling history, i.e., by additionally writing the values of $S_1, S_2, ..., S_{N-1}$ to those of S_N and m_N. A single qubit state has $S_1 = \frac{1}{2}$, a two-qubit system can either have $S_2 = 0$ or $S_2 = 1$, a three-qubit system $S_3 = \frac{1}{2}$ or $S_3 = \frac{3}{2}$, and so on. Including the coupling history we thus arrive at the notation $|S_1, S_2, ..., S_N; m_N\rangle$ which describes a particular angular momentum eigenstate unambiguously.

3.4. GENERATION OF TAME IN REMOTE QUBITS

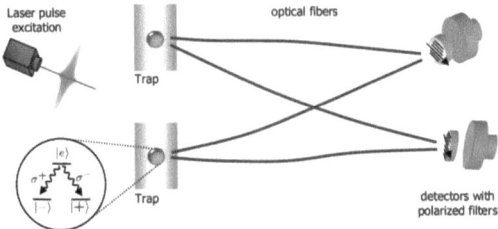

Figure 3.7: Experimental setup for the angular momentum coupling of two atoms via projective measurements using optical fibers. In a successful measurement cycle, each atom emits a single photon and each detector registers exactly one photon. Note that the detectors cannot distinguish which of the atoms emitted a registered photon.

3.4.3 Setup

In the following, we consider a system of N trapped atoms with a Λ-configuration. The two ground states $|+\rangle$ and $|-\rangle$ of the Λ-scheme can be expressed, using the notation introduced before, as $|\frac{1}{2};+\frac{1}{2}\rangle \equiv |+\rangle$ and $|\frac{1}{2};-\frac{1}{2}\rangle \equiv |-\rangle$. Initially, all atoms are excited by a laser π pulse towards the excited state $|e\rangle$ and subsequently decay by spontaneously emitting N photons that are collected by single-mode optical fibers and transmitted to N different detectors. Since each atom is connected via optical fibers to several detectors, a single photon can travel on several alternative, yet equally probable paths to be eventually recorded by one detector (cf. Sec. 3.2.4). After a successful measurement in which all N photons have been recorded at the N detectors so that each detector registers exactly one photon, it is thus impossible to determine along *which way* each of the N photons propagated. As described, this may cause quantum interferences of Nth order which can be fruitfully employed to engineer particular quantum states of the emitters, e.g., to generate families of entangled states symmetric under permutation of their qubits [81, 82]. Here, we will consider the generation of a more general class of quantum states, including symmetric *and* nonsymmetric states. By mimicking the process of spin-spin coupling, we will demonstrate how to generate any quantum state belonging to the coupled basis of an N-qubit compound system.

3.4.4 Measurement-based Preparation of Total Angular Momentum Eigenstates

Let us start by looking at the most basic process of our system. In Sec. 3.2.2 we saw that one single excited atom with a Λ-configuration emits a photon, the atomic ground state and the photonic polarization states cannot be described independently. The excited state $|e\rangle$ can decay along two possible channels, $|e\rangle \to |+\rangle$ and $|e\rangle \to |-\rangle$, accompanied by the spontaneous emission of a σ^-- or a σ^+-polarized photon, respectively.

A single decaying atom thus forms an entangled state between the polarization state of the emitted photon and the corresponding ground state of the de-excited atom. This correlation implies that the state of the atom is projected onto its total angular momentum eigenstate $|+\rangle$ ($|-\rangle$) if the emitted photon is registered by a detector equipped with a σ^-- (σ^+-) polarized filter (cf. Sec. 3.2.3). Since these are the only two TAMEs that exist for a single qubit, this simple method already allows to create all TAMEs for a single qubit.

Preparation of 2-qubit states

In a next step, we consider the system shown in Fig. 3.7 where two atoms with a Λ-configuration are initially excited and the spontaneously emitted photons are subsequently measured at two different detectors. Again, if a polarization sensitive measurement is performed on the emitted photons using two different polarization filters in front of the detectors, the state of the two atoms is projected due to the measurement (cf. Sec. 3.2.4). In the particular case that the polarization of both photons is measured along orthogonal directions, the state of the atoms will be projected onto a superposition of both ground states, since it is impossible to determine which atom emitted the photon travelling to the first or the second detector by the information obtained in the measurement process. Note that with each qubit having a total spin of $\frac{1}{2}$, a two-qubit system can have a total spin of either 1 or 0 so that the four total angular momentum eigenstates are given by

$$\begin{array}{cccc}
\text{spin-1 triplet} & |S_1,S_2;m\rangle & \text{spin-0 singlet} & |S_1,S_2;m\rangle \\
|++\rangle & |\tfrac{1}{2},1;+1\rangle & & \\
\tfrac{1}{\sqrt{2}}(|+-\rangle+|-+\rangle) & |\tfrac{1}{2},1;0\rangle & \tfrac{1}{\sqrt{2}}(|+-\rangle-|-+\rangle) & |\tfrac{1}{2},0;0\rangle \\
|--\rangle & |\tfrac{1}{2},1;-1\rangle & &
\end{array}$$

The spin-1 triplet can be easily generated with the setup shown in Fig. 3.7 by fixing the length of the optical fibers such that the phase accumulated by a photon in each fiber is $\varphi = 2\pi n$ ($n \in \mathbb{N}$) and choosing the polarization filters appropriately: For example, if both filters are oriented in such a way that only σ^-- (σ^+-) polarized photons are transmitted, the emitters are projected onto the state $|++\rangle$ ($|--\rangle$); if the filters are orthogonal, i.e., one is transmitting σ^-- and one σ^+-polarized photons, the system is projected onto the state $|\tfrac{1}{2},1;0\rangle$, since any information along *which way* the photons propagated is erased by the system. Finally, in order to generate the singlet state $|\tfrac{1}{2},0;0\rangle$, we may introduce an additional optical phase shift of π in only one of the optical fibers shown in Fig. 3.7, e.g., by extending or shortening the length of the optical path by $\tfrac{\lambda}{2}$. The generation of the four two-particle TAMEs with the system

3.4. GENERATION OF TAME IN REMOTE QUBITS

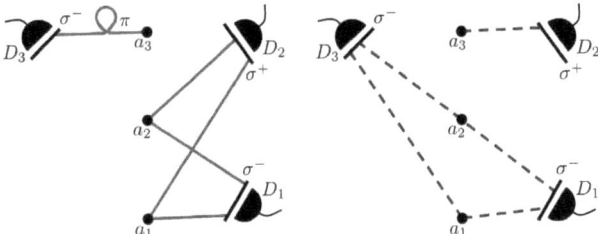

Figure 3.8: Left: Extension of the setup shown in Fig. 3.7 capable of generating the state $|\frac{1}{2},1;0\rangle \otimes |+\rangle$. Right: configuration for the generation of the state $|\frac{1}{2},1;1\rangle \otimes |-\rangle$.

shown in Fig. 3.7 thus requires only the variation of two polarizer orientations and, in case of the singlet state, to introduce an optical phase shift of π.

Preparation of 3-qubit states

With the two-qubit angular momentum eigenstates at hand, we can next couple an additional qubit in order to access the eight possible three-qubit total angular momentum eigenstates. In the following, we will exemplify our method for the non-symmetric three-qubit state $|\frac{1}{2},1,\frac{1}{2};+\frac{1}{2}\rangle$ given by

$$\begin{aligned}|\frac{1}{2},1,\frac{1}{2};+\frac{1}{2}\rangle &= \frac{1}{\sqrt{6}}(2|++-\rangle - |+-+\rangle - |-++\rangle) \\ &= \frac{1}{\sqrt{6}}(2|++\rangle \otimes |-\rangle - (|+-\rangle - |-+\rangle) \otimes |+\rangle) \\ &= \frac{\sqrt{2}}{\sqrt{3}}|\frac{1}{2},1;+1\rangle \otimes |-\rangle - \frac{1}{\sqrt{3}}|\frac{1}{2},1;0\rangle \otimes |+\rangle,\end{aligned} \quad (3.28)$$

where the last line in Eq. (3.28) exhibits the coupling history: In order to generate the three-qubit state $|\frac{1}{2},1,\frac{1}{2};+\frac{1}{2}\rangle$, the two-qubit spin-1 states $|\frac{1}{2},1;+1\rangle$ and $|\frac{1}{2},1;0\rangle$ have to be appropriately coupled with the basis states $|-\rangle$ and $|+\rangle$ of a third qubit. Thereby, the prefactors $\frac{\sqrt{2}}{\sqrt{3}}$ and $-\frac{\sqrt{1}}{\sqrt{3}}$ represent the corresponding Clebsch-Gordan coefficients as a result of changing the basis [100]. In the following, we will make use of our knowledge of how to generate the states $|\frac{1}{2},1;+1\rangle$ and $|\frac{1}{2},1;0\rangle$ in order to generate the desired state $|\frac{1}{2},1,\frac{1}{2};+\frac{1}{2}\rangle$.

The two setups individually capable of generating the three-qubit states $|\frac{1}{2},1;+1\rangle \otimes |-\rangle$ and $|\frac{1}{2},1;0\rangle \otimes |+\rangle$ are shown in Fig. 3.8. The additional qubit is not yet coupled to the two-qubit system, i.e., it is simply projected either onto the state $|+\rangle$ (Fig. 3.8, left) or $|-\rangle$ (Fig. 3.8, right), where the two-qubit systems are projected in the same way as explained in Fig. 3.7. In order to generate the three-qubit state $|\frac{1}{2},1,\frac{1}{2};\frac{1}{2}\rangle$, we now have

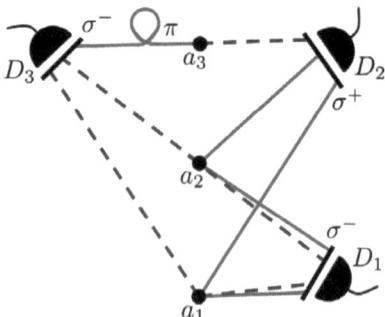

Figure 3.9: Setup for the generation of the state $2|++-\rangle - |+-+\rangle - |-++\rangle$. The blue dashed lines indicate the quantum path which leads to $2|++-\rangle$, whereas the red solid labeled path indicate the quantum paths which lead to $-|+-+\rangle - |-++\rangle$. The different red solid and blue dashed lines leading from atom a_1 (a_2) to detector D_1 are drawn to indicate the different quantum paths only. Physically, there is only one fiber from atom a_1 (a_2) to detector D_1.

to superpose these two possibilities. The combined system is shown in Fig. 3.9. We will explain the underlying physics by considering the possible scenarios when detecting the photon emitted by the additional third atom.

In a successful measurement cycle, the three emitted photons are detected at three different detectors. Thus, there are only two possible situations with regard to a measurement of a photon emitted by the third atom:

I. (red solid lines) The emitted photon is registered at detector D_3 which has a σ^- polarizing filter in front. In this case, emitter a_3 is projected onto the state $|+\rangle$ and emitter a_1 and a_2 are left in the setup generating the state $|\frac{1}{2},1;0\rangle \equiv \frac{1}{\sqrt{2}}(|+-\rangle + |-+\rangle)$, as discussed in Fig. 3.8 (left).

II. (blue dashed lines) The emitted photon is registered at detector D_2 which has a σ^+ polarizing filter in front. In this case, emitter a_3 is projected onto the state $|-\rangle$ and emitter a_1 and a_2 are left in the setup generating the state $|\frac{1}{2},1;1\rangle \equiv |++\rangle$, as discussed in Fig. 3.8 (right).

In other words, the third emitter acts as a switch between the two possible quantum paths: with equal probabilities, the system is either projected onto the state $2|++-\rangle$ or onto the state $|+-+\rangle + |-++\rangle$. Since both possibilities exist at the same time and we do not know which of the two will be realized, we have to write the corresponding quantum state as a superposition of these two possibilities. Note that the relative factor of two results from the two possibilities of obtaining the state $|++-\rangle$: either a_1 emits a

3.4. GENERATION OF TAME IN REMOTE QUBITS

photon which is detected at D_1 and a_2 sends a photon to D_2 or vice versa. In addition, we can modify the path where a photon emitted by the third atom is registered at detector D_3 by implementing a relative optical phase shift of π (cf. Fig. 3.9) to obtain a minus sign for scenario II relative to scenario I. In this case, the final state projected by the setup shown in Fig. 3.9 corresponds to the three-qubit state $|\frac{1}{2},1,\frac{1}{2};\frac{1}{2}\rangle$ of Eq. (3.28).

Reconsidering the state $|\frac{1}{2},1,\frac{1}{2};\frac{1}{2}\rangle$ in terms of our extended notation, we coupled two spin-1/2 particles to form a spin-1 compound state that was coupled again with a spin-1/2 particle to form a three-particle spin-1/2 compound state. Similarly, we could have coupled the spin-1 compound state with an additional qubit in such a way that we obtain the symmetric state $|\frac{1}{2},1,\frac{3}{2};\frac{1}{2}\rangle$, also known as W-state [96]. For this case, we have to change the setup shown in Fig. 3.9 slightly: we remove the optical phase shift of π and connect the third emitter also with detector D_1. In this case, the totally symmetric setup generates a W-state (cf. [81]).

Preparation of N-qubit states

Following the scenario of the preceeding section, we can now outline how to engineer the coupling of angular momentum of N remote qubits to form an arbitrary N-qubit total angular momentum eigenstate. In order to generate the N-qubit state $|S_1, S_2, S_3, ...S_N; m_s\rangle$ we have to

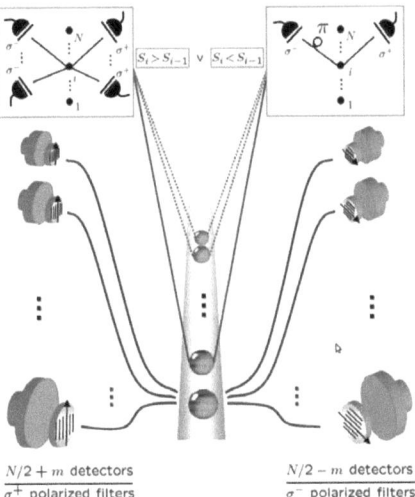

Figure 3.10: Experimental setup for the spin-spin coupling of N remote atoms via projective measurements.

1. set up $\frac{N}{2}+m_s$ ($\frac{N}{2}-m_s$) detectors with σ^-- (σ^+-) polarized filters in front. Hereby, we connect the first emitter with optical fibers to all N detectors.

2. check for each particle i beginning with $i = 2$ whether $S_i > S_{i-1}$ or $S_i < S_{i-1}$. If

   a. $S_i > S_{i-1}$; we have to connect the particle with optical fibers to all detectors except those which are mentioned in case b. below.

   b. $S_i < S_{i-1}$; we have to connect the particle with optical fibers to one detector with a σ^- polarizer and to one with a σ^+ polarizer. The optical fiber leading to the σ^- polarizer should induce a relative optical phase shift of π and those two detectors should not be linked to any other subsequent particle.

If one wants to create a particular TAME $|S_1, S_2, S_3, ...S_N; m_s\rangle$, the setup is uniquely determined by the total spins $S_1, S_2, S_3, ...S_N$ obtained by successively coupling N spin-1/2 particles. Hereby, the spin quantum number m_s determines the fraction of σ^- and σ^+ polarized filters used in the setup (see Fig. 3.10). While this algorithm has been derived from the case $N = 3, 4$ where the physical picture tells us that it should also work for higher N, the mathematical proof that this is indeed the case is far from trivial. It relies on the idea of possibly populating N modes by a photon emitted from one atom, i.e., the formalism of the population operator (cf. Sec. 3.7) and its inductive application. For details, the interested reader is referred to [101].

3.4.5 Examples

As examples, let us apply this algorithm for the two three-qubit total angular momentum eigenstates $|\frac{1}{2},1,\frac{1}{2};\frac{1}{2}\rangle$ and $|\frac{1}{2},1,\frac{3}{2};\frac{1}{2}\rangle$ discussed above. Since $m_s = \frac{1}{2}$ for both states, we use two detectors with σ^--polarized filters and one with a σ^+-polarized filter. Further, in both cases we have $S_2 > S_1$ which implies that the first and the second emitter are connected to all three detectors. For the state $|\frac{1}{2},1,\frac{1}{2};\frac{1}{2}\rangle$, we find $S_3 < S_2$. Therefore, we connect the third emitter only to two detectors with σ^-- and σ^+-polarized filters in front, respectively, e.g. detector D_2 and D_3, and we introduce an optical phase shift of π for the path leading from the third emitter to detector D_3. Summarizing we obtain the setup shown in Fig. 3.9 as postulated. For the state $|\frac{1}{2},1,\frac{3}{2};\frac{1}{2}\rangle$, we find $S_3 > S_2$. Here, we connect the third emitter to all three detectors and do not introduce an additional phase shift anywhere. In this case, as mentioned above, the setup will generate the symmetric W-state [81].

After the setup has been constructed according to the algorithm, one can also take a look at the detection operators which follow from the setup. We will again take the state $|\frac{1}{2},1,\frac{1}{2};\frac{1}{2}\rangle$ as an example. Following Fig. 3.9, the detection operators can be

3.4. GENERATION OF TAME IN REMOTE QUBITS

written as

$$\hat{D}_1 = |+\rangle_{a_1}\langle e| + |+\rangle_{a_2}\langle e|, \qquad \hat{D}_2 = |-\rangle_{a_1}\langle e| + |-\rangle_{a_2}\langle e| + |-\rangle_{a_3}\langle e|, \text{ and}$$
$$\hat{D}_3 = |+\rangle_{a_1}\langle e| + |+\rangle_{a_2}\langle e| - |+\rangle_{a_3}\langle e|. \qquad (3.29)$$

Unfortunately, it is not possible to translate the algorithm given above into an algorithm which describes the detailed form of the detection operators.[5]

It should also be mentioned that in order to generate a specific TAME, there exist more possibilities of setting up the detectors than the one prescribed by the algorithm. For example, the state $|\frac{1}{2},1,\frac{1}{2};\frac{1}{2}\rangle$ might as well be generated by use of the detection operators:

$$\hat{D}_{1'} = |+\rangle_{a_1}\langle e| + |+\rangle_{a_2}\langle e| + |+\rangle_{a_3}\langle e|, \qquad \hat{D}_{2'} = |+\rangle_{a_1}\langle e| + |+\rangle_{a_2}\langle e| + |+\rangle_{a_3}\langle e|, \text{ and}$$
$$\hat{D}_{3'} = |-\rangle_{a_3}\langle e| - \tfrac{1}{2}|-\rangle_{a_1}\langle e| - \tfrac{1}{2}|-\rangle_{a_2}\langle e|, \qquad (3.30)$$

where a phase shift of π is introduced for the two fibers leading from atoms a_1 and a_2 to D_3 and the prefactor of $\frac{1}{2}$ can be realized by attenuating the output of the fibers accordingly (cf. Sec. 3.6.2).

3.4.6 Conclusion

The method proposed here relies on the probabilistic scattering of photons. Thereby, the efficiency of generating a particular N-qubit total angular momentum eigenstate decreases with increasing number of qubits N. If the probability to find a single photon in an angular detection window $\Delta\Omega$ is given by $P(\Delta\Omega)$, including fiber coupling and detection efficiencies, the corresponding N-fold counting rate is found to be proportional to $P^N(\Delta\Omega)$. This might limit the scalability of our scheme (see the discussion in [81]) as is indeed the case with other experiments observing entangled atoms [83, 86, 87].

We considered a system of N remote noninteracting single-photon emitters with a Λ-configuration. By mimicking the coupling of angular momentum, we showed that it is possible to engineer any of the 2^N total angular momentum eigenstates in the long-lived ground-state qubits. Using linear optical tools only, our method employs the detection of all N photons scattered from the N emitters at N polarization sensitive detectors. Thereby, it offers access to any of the 2^N states of the coupled basis of an N-qubit compound system. Using projective measurements we thereby form highly and weakly entangled quantum states even though no interaction between the qubits is present.

---

[5]This is the case because the algorithm is formulated from the "viewpoint" of the atoms while the detection operators describe the situation from the "viewpoint" of the detectors.

3.5 Generation of Cluster States in Remote Qubits

3.5.1 Introduction

In 2001, Raussendorf and Briegel suggested that a quantum computer can be realized by simple local measurements only, if these measurements are conducted on a certain family of entangled states, the so-called cluster states [56]. This, however, only shifted the difficulties from the implementation of quantum gates to the generation of a cluster state. So far, numerous proposals have been put forth of how to generate cluster states in different physical systems [27, 102–123]. Some of the detection based schemes are also implementable in our setup, as we will show below on the basis of [120].

3.5.2 A Proposal for Cluster States Generation by Xia et al.

In a recent paper, Xia et al. proposed a setup for the generation of a class of cluster states in Λ-type atoms of $K = M + N$ qubits based on projection via detection of emitted photons [120]. Their setup is depicted in Fig. 3.11.

The scheme can be described in the following way: first, quarter-wave plates (QWP) transform the σ^-- and σ^+-polarized photons emitted by the atoms into horizontally and vertically polarised photons, respectively. The (non-normalized[6]) state of the system after the QWPs is given by

$$|\psi^{(i)}\rangle = \bigotimes_{i=\text{all atoms}} (|+\rangle_i |H\rangle_i + |-\rangle_i |V\rangle_i). \quad (3.31)$$

Thereafter, polarizing beam splitters (PBSs) which are supposed to transmit horizontally polarized photons and reflect vertically polarized photons transform the system into a state of the form (there is an error in Eq. (2) of the original paper)

$$|\psi'\rangle = \left[(|+\rangle_{A_1}|H\rangle_{b_M} + |-\rangle_{A_1}|V\rangle_{b_1}) \otimes \bigotimes_{i=2,\ldots,M} (|+\rangle_{A_i}|H\rangle_{b_{i-1}} + |-\rangle_{A_i}|V\rangle_{b_i}) \right] \otimes \\ \left[(|+\rangle_{B_1}|H\rangle_N + |-\rangle_{B_1}|V\rangle_1) \otimes \bigotimes_{i=2,\ldots,N} (|+\rangle_{B_i}|H\rangle_{i-1} + |-\rangle_{B_i}|V\rangle_i) \right], \quad (3.32)$$

where the indices A_i, b_i ($i = 1, \ldots, M$) and B_i, i ($i = 1, \ldots, N$) refer to the emitters and modes as indicated in Fig. 3.11. In a postselection process, we restrict ourselves to the cases where all the modes 1,...,N and $b_1, ..., b_M$ contain exactly one photon. Hence, discarding all parts of the state vector describing bunched outcomes from Eq. (3.32),

---
[6]Normalization factors will always be omitted throughout this section.

3.5. GENERATION OF CLUSTER STATES

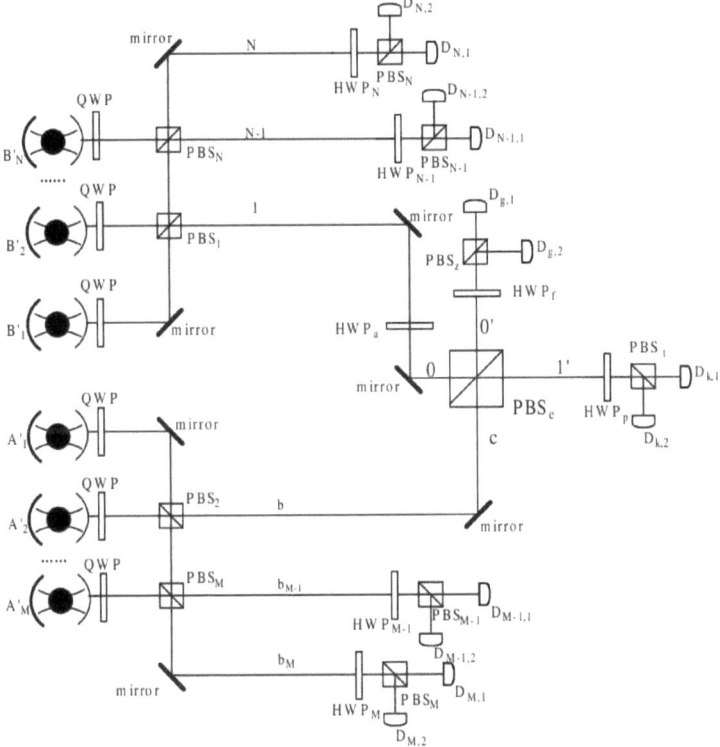

Figure 3.11: Beam splitter based setup to generate cluster states in atomic qubits (Figure from [120]).

the wavefunction can be written as

$$|\psi''\rangle = \left[\bigotimes_{i=1,...,M} |+\rangle_{A_i}|H\rangle_{b_i} + \bigotimes_{i=1,...,M} |-\rangle_{A_i}|V\rangle_{b_i}\right] \otimes$$
$$\left[\bigotimes_{i=1,...,N} |+\rangle_{B_i}|H\rangle_i + \bigotimes_{i=1,...,N} |-\rangle_{B_i}|V\rangle_i\right]$$
$$= (|+\rangle|H\rangle)^{M+N} + |+\rangle^M|-\rangle^N|H\rangle^M|V\rangle^N + |-\rangle^M|+\rangle^N|V\rangle^M|H\rangle^N + (|-\rangle|V\rangle)^{M+N}. \tag{3.33}$$

This state is very close to a cluster state, implemented in the atomic as well is in the photonic degrees of freedom. Only a phase of π is missing in the last term. Now the photon in mode 1 will meet a half-wave plate (HWP) aligned such that it realizes the transformations $|H\rangle \to \frac{1}{\sqrt{2}}(|H\rangle + |V\rangle)$ and $|V\rangle \to \frac{1}{\sqrt{2}}(|H\rangle - |V\rangle)$, which will further transform the state of Eq. (3.33) into

$$|\psi'''\rangle = \left[\bigotimes_{i=1,\dots,M} |+\rangle_{A_i} |H\rangle_{b_i} + \bigotimes_{i=1,\dots,M} |-\rangle_{A_i} |V\rangle_{b_i}\right] \otimes$$

$$\left[\left(|+\rangle_{B_1}(|H\rangle_1) + |V\rangle_1)\bigotimes_{i=2,\dots,N} |+\rangle_{B_i}|H\rangle_i\right)\right.$$

$$\left. + \left(|-\rangle_{B_1}(|H\rangle_1) - |V\rangle_1)\bigotimes_{i=2,\dots,N} |-\rangle_{B_i}|V\rangle_i\right)\right]. \quad (3.34)$$

Again discarding photon bunched outcomes in the modes $0'$ and $1'$ after the polarizing beam splitter PBS_e, the state reads

$$|\psi''''\rangle = \bigotimes_{\text{all atoms}} |+\rangle \bigotimes_{\text{all modes}} |H\rangle$$

$$+ \left(\bigotimes_{i=1,\dots,M} |+\rangle_{A_i}\right)\left(\bigotimes_{i=1,\dots,N} |-\rangle_{B_i}\right) |H\rangle_{0'}|H\rangle_{b_2}|H\rangle_{b_3}\dots|H\rangle_{b_M}|H\rangle_{1'}|V\rangle_2|V\rangle_3\dots|V\rangle_N$$

$$+ \left(\bigotimes_{i=1,\dots,M} |-\rangle_{A_i}\right)\left(\bigotimes_{i=1,\dots,N} |+\rangle_{B_i}\right) |V\rangle_{0'}|V\rangle_{b_2}|V\rangle_{b_3}\dots|V\rangle_{b_M}|V\rangle_{1'}|H\rangle_2|H\rangle_3\dots|H\rangle_N$$

$$- \bigotimes_{\text{all atoms}} |-\rangle \bigotimes_{\text{all modes}} |V\rangle. \quad (3.35)$$

This is a cluster state in atoms and photons. By measuring the photons in the basis $\{(|H\rangle + |V\rangle), (|H\rangle - |V\rangle)\}$, achieved by the HWPs and PBSs in front of the detectors, the atoms are projected into one of the four cluster states

$$|\psi\rangle = \begin{cases} \frac{1}{2}\left(|+\rangle^M|+\rangle^N + |+\rangle^M|-\rangle^N + |-\rangle^M|+\rangle^N - |-\rangle^M|-\rangle^N\right) \\ \frac{1}{2}\left(|+\rangle^M|+\rangle^N + |+\rangle^M|-\rangle^N - |-\rangle^M|+\rangle^N + |-\rangle^M|-\rangle^N\right) \\ \frac{1}{2}\left(|+\rangle^M|+\rangle^N - |+\rangle^M|-\rangle^N + |-\rangle^M|+\rangle^N + |-\rangle^M|-\rangle^N\right) \\ \frac{1}{2}\left(|+\rangle^M|+\rangle^N - |+\rangle^M|-\rangle^N - |-\rangle^M|+\rangle^N - |-\rangle^M|-\rangle^N\right) \end{cases}, \quad (3.36)$$

which are all equivalent to each other up to the local unitary transformation of exchanging $|+\rangle$ and $|-\rangle$ in all A or all B-atoms. It depends on the combination of basis states that is observed, i.e., on the combination of detectors that actually click, which of th four possible cluster states is actually realized.

3.5. GENERATION OF CLUSTER STATES

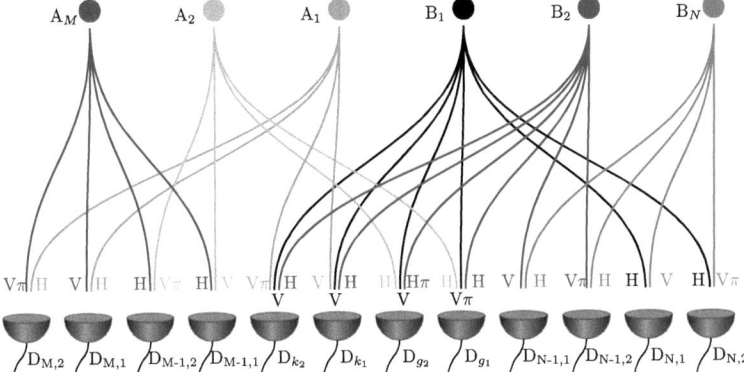

Figure 3.12: The fiber-based setup, in analogy to Fig. 3.11. H and V describe, which polarization the fiber transmits and a π indicates a phase shift of π with respect to the other fibers. A QWP which transforms $|\sigma^{\pm}\rangle$ to $|{}^V_H\rangle$ placed after every atom is assumed to be placed between every atom and the fibers connected to it. To obtain an uncrowded sketch, these QWPs are not drawn.

3.5.3 Translating the Proposal into a Fiber-based Setup

In order to translate this scheme into our setup of Λ-level atoms emitting photons recorded by detectors such that the origin of the source remains unknown, one needs to identify every possible optical path in the setup from Fig. 3.11 and realize that path with a fiber. For example, in the scheme by Xia et al., a σ^--polarized photon emitted by atom A_M will be transformed into a vertically polarized one and consequently be reflected at the polarizing beam splitter PBS_M. At the half-wave plate, it either remains a vertically polarized photon, picks up a phase of π and is detected at $D_{M,2}$ or it is transformed into a horizontally polarized photon and recorded by $D_{M,1}$. By contrast, an initially σ^+-polarized photon emitted by atom A_M never arrives at either of the two detectors because it will be deflected into a different mode by the first polarizing beam splitter. Thus, in a fiber-based approach, there should be one fiber with a π phase shift and a V polarization filter at the end of the fiber leading from the QWP after atom A_M to detector $D_{M,1}$ and a second fiber without a phase shift and with the same V polarization filter at its end from atom A_M to $D_{M,2}$. If this is done for all optical paths which are realized in Fig. 3.11, we arrive at the scheme depicted in Fig. 3.12.

We note that in contrast to the schemes we have developed in the previous sections, we need a polarization filter after every fiber, even if the fibers end at the same detector. We note further that there are $2N$ detectors, i.e., twice as many detectors as photons

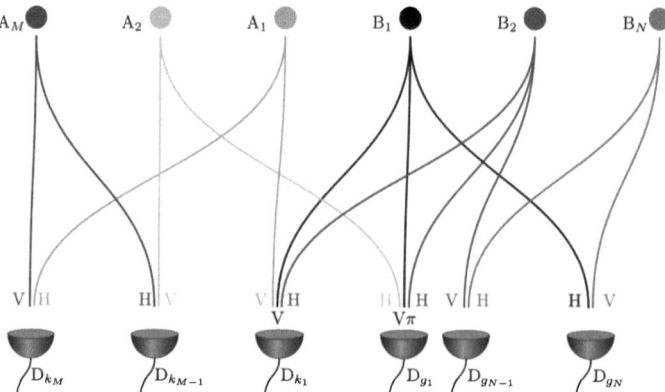

Figure 3.13: The simplified fiber-based setup, where only six arbitrarily chosen detectors remain in the setup.

in the setup, while only N detectors must click in order to project the atoms: We have seen in Sec. 3.5.2 that (up to a local unitary transformation) the same cluster state is generated if and only if one of the two detectors connected to each of the modes $0', 1', 2, \ldots, N, b_2, \ldots, b_M$ in Fig. 3.11 clicks. Thus, we may as well restrict the setup from Fig. 3.12 in an arbitrary choice to only those detectors, where the last index is a 1. In that case, the setup simplifies to the one depicted in Fig. 3.13. The detection operators which correspond to this detector setup have the form

$$\begin{aligned}
\hat{D}_{k_1} &= |-\rangle_{A_m}\langle e| + |-\rangle_{B_1}\langle e| + |+\rangle_{B_2}\langle e|, & \hat{D}_{g_1} &= |+\rangle_{A_2}\langle e| - |-\rangle_{B_1}\langle e| + |+\rangle_{B_2}\langle e|, \\
\hat{D}_{k_i} &= |-\rangle_{A_i}\langle e| + |+\rangle_{A_{i+1}}\langle e|, & \hat{D}_{g_j} &= |-\rangle_{B_j}\langle e| + |+\rangle_{B_{j+1}}\langle e|, \quad (3.37) \\
\hat{D}_{k_M} &= |-\rangle_{A_M}\langle e| + |+\rangle_{A_1}\langle e|, \text{ and} & \hat{D}_{g_N} &= |-\rangle_{B_N}\langle e| + |+\rangle_{B_1}\langle e|
\end{aligned}$$

with $i \in \{2, 3, \ldots, M-1\}$ and $j \in \{2, 3, \ldots, N-1\}$.

Note that the sketch is oversimplified: the photons at the end of the fibers are not principally indistinguishable and thus would not interfere at the detectors. We first need to make them indistinguishable, for example by using an additional linear polarization filter in front of the detector, oriented halfway between H and V. However, this diminishes the detection probability, since there is an additional 50 % probability for each photon to be absorbed at this filter. Thus, we choose a different method to erase the WW information: we insert a half-wave plate into all fibers that are supposed to transmit only V-polarized light. The half-wave plate is oriented at $\pi/4$, such that a V-polarized photon is transformed into an H-polarized photon (and vice versa). In

3.5. GENERATION OF CLUSTER STATES

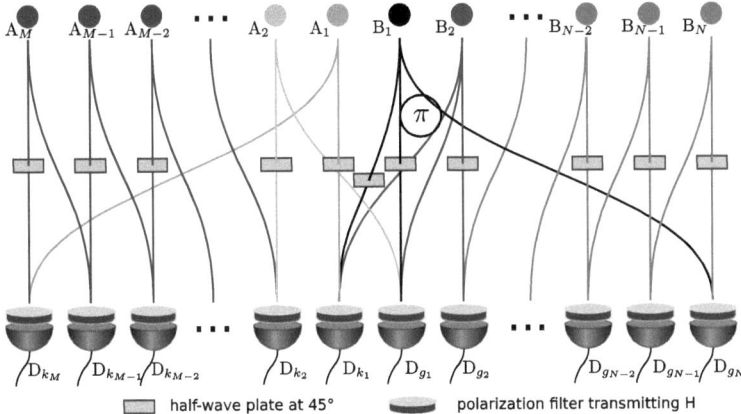

Figure 3.14: The improved fiber scheme. It is drawn for an arbitrary number of atoms in order to demonstrate the physics behind the algorithm: consider the case that atom A_M emitted a V-polarized photon. Because of our post-selection condition that every photon is detected, the V-photon must have travelled to detector D_{k_M}. Since detector $D_{k_{M-1}}$ is only connected to A_M and A_{M-1}, we find that A_{M-1} must have emitted to that detector, thus atom A_{M-1} must be in the same state as A_M. Following this logic through, we find that all A-atoms must be in the same state if all detectors registered a photon. The same is true for the B-atoms: the state of one is arbitrary, but all others must be in the same state. This leaves us with the four only states $|+\rangle_A|+\rangle_B$, $|+\rangle_A|-\rangle_B$, $|-\rangle_A|+\rangle_B$, and $|-\rangle_A|-\rangle_B$, which can contribute to our postselected events. The phases between these states are found by noting that only in front of $|-\rangle_A|-\rangle_B$ the fiber with the π phase shift contributes.

that setup, we also need an H-polarizer (instead of a V-polarizer) after the fibers that are supposed to transmit only photons that were originally H-polarized. Since we have only H-oriented polarization filters, we can replace the seperate polarization filters after all fibers leading to a certain detector by a single H-oriented polarization filter after the beams of the fibers have been recombined. The corresponding setup is shown in Fig. 3.14. As we will see, this change in the setup will enable us to generate photonic cluster states (cf. Sec. 3.7).

Assuming perfect components (no losses, 100 % detection probability) the probability of finding exactly one photon in each mode is given by

$$P(N) = \frac{1}{3^2 2^{N+M-2}}. \quad (3.38)$$

However, every (initially unpolarized) photon also has to pass a polarization filter which happens with a probability of $\frac{1}{2}$ at every polarization filter, so that the total success probability of the scheme is given by

$$P_{\text{fib}}(N) = \left(\frac{1}{2^{N+M}}\right) P(N). \tag{3.39}$$

The additional lowering of the success probability due to the polarization filters could be overcome by using polarizing beam splitters behind the QWP after each atom to split the modes into the two or three different fibers. This would eliminate the need for polarization filters and the success probability would be given by Eq. (3.38). However, the translation of this scheme into one for the generation of photonic cluster states is not straightforward anymore (cf. Sec. 3.7), therefore, it is not discussed in more detail here.

3.5.4 Conclusion

In this section, we have shown that it is in principle possible to create a cluster state with an arbitrary number of qubits by detecting the scattered photons. However, this method does not make use of one of the most remarkable properties of the cluster state, namely that it can be build up by *sequentially* entangling Bell pairs with the already existing cluster state [124]. Protocols which make use of this property (e.g. [125]) have a success probability which scales only polynomially instead of exponentially and might therefore in general be better suited in an actual implementation.

3.6 Generation of Arbitrary States: The Anystate?

So far, we have seen that the method of projective measurements is capable of producing an impressive number of states. Naturally, the question arises whether it is possible to create any arbitrary state (from now on termed the anystate) of N qubits by arranging the detectors whitfully. In the following, we will solve this problem explicitly for $N = 2, 3$. In addition, we will argue that already simple considerations of the number of degrees of freedom involved in the system show that it cannot be possible to generate the anystate for $N \geq 7$.

3.6.1 Fundamentals

The most general quantum mechanical pure state of N qubits is given by

$$|A_N\rangle = \sum_{j=0}^{2^N-1} c_j |j\rangle, \qquad (3.40)$$

where the state $|j\rangle$ is determined by the binary form if one identifies each $+$ with a 1 and each $-$ with a 0, e.g., $|+-+\rangle = |101\rangle = |5\rangle$. The prefactors c_n are arbitrary complex numbers with the restriction of normalization $\sum |c_j|^2 = 1$. This is the anystate, which we want to generate by the detection of N photons having passed an array of appropriate linear optical elements. The final state of the atoms is calculated by applying N detection operators to the initial wavefunction where all atoms are in the excited state $|ee\ldots e\rangle$. As before, it only affects the count rate if the excitation mechanism is non-perfect, because only the fully excited part of the wavefunction contributes to the Nth-order correlation function. The most general form of the detection operator for a photon detected at detector D_i (cf. Eq. (3.7)) is given by

$$\hat{D}_i = \sum_{j=1}^{N} \left(d_{+ji} |+\rangle_{a_j}\langle e| + d_{-ji} |-\rangle_{a_j}\langle e| \right). \qquad (3.41)$$

If we consider every single parameter $d_{\pm j}$ as independent, then there are $2N$ complex degrees of freedom that can be adjusted for every one of the N detectors, i.e., a total of $2N^2$ variables. However, the number of prefactors in the most general state of N qubits grows with 2^N (see Eq. (3.40)). This number surpasses the one of the free variables already for $N = 7$ (98 free variables for 128 prefactors). Even allowing for local unitary transformations on every qubit after all photons have been detected would not help to go to higher N because in this case only another N degrees of freedom are additionally available. From this dimensional argument, one can already see that it will not be possible to create the anystate for $N \geq 7$. Before we discuss whether the creation of the anystate is possible for $N < 7$, let us take a look at how the most general detection

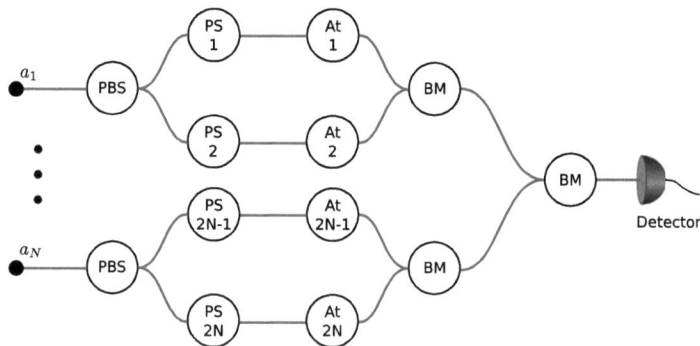

Figure 3.15: Schematic drawing of the setup how to link N sources to one particular detector D_i in order to realize the most general detection operator $\hat{D}_i$ from Eq. (3.41): the light coming from each photon is split at a polarizing beam splitter (PBS) into its σ^-- and σ^+-components, then each component passes a phase shifter (PS) and an attenuater (At). Thereafter, all beams are combined with beam mergers (BM).

operator might be implemented experimentally.

3.6.2 Physical Realization of an Arbitrary Detection Operator

In order to realize the most general detection operator physically, we must gain operational control of each prefactor $d_{\pm j}$ in Eq. (3.41) seperately. This criterion already excludes far field detection schemes, as they neither allow for only for different polarization of the photons emitted from any of the N atoms and detected at a given detector D_i nor do they give access to all phases individually. However, we can realize every summand of Eq. (3.41) separately by guiding the photons through the setup depicted in Fig. 3.15 which realizes a spatial seperation of all quantum paths that a photon which is detected at D_i might have taken.

In this setup, the photon path from each atom leads through a polarizing beam splitter (PBS) which splits the beam into its components of σ^+- and σ^--polarization. After the N PBSs behind the N atoms we have $2N$ beams, each corresponding exactly to one of the summands in Eq. (3.41). With the phase shifters (PS) and the attenuators (At), we have operational control over the phase and the modulus of each prefactor of Eq. (3.41) seperately.[7] Of course, it is not possible to amplify the signals, and we are thus restricted in the absolut value of the prefactors to the ones of the unattenuated

---

[7]Actually, we could remove one phase shifter and one attenuater, since only the relative phases and the relative intensities are of importance. However, the symmetric setup of Fig. 3.15 is more transparent, thus we will stick to this picture.

3.6. GENERATION OF ARBITRARY STATES: THE ANYSTATE?

beams. However, as mentioned before, only the relative intensities are of importance and we can thus take the unattenuated beam as reference and scale down all other beams appropriately.

3.6.3 Generating the Anystate N=2

In Sec. 3.3 we have already calculated the general final state of two Λ-level atoms if the photons they emitted were allowed to propagate in free space and hit polarization sensitive detectors. It is easy to show that Eq. (3.13) cannot be the anystate in disguise by a simple counterexample: the state $|+-\rangle$ is impossible to create because one has to get rid of the contributions of $|++\rangle$ and $|--\rangle$. This would imply that either at least one of their respective prefactors $\varepsilon_{(1,2)-}$ and $\varepsilon_{(1,2)+}$ vanishes or that $e^{i\varphi_{21}} = -1$. However, from either of the two requirements one can conclude that the prefactors of $|+-\rangle$ and $|-+\rangle$ differ at the most by a complex phase but not in their amplitude. Consequently, $|+-\rangle$ cannot be generated by the setup because it would require an amplitude of 1 for $|+-\rangle$ and an amplitude of 0 for $|-+\rangle$. Therefore, already for $N = 2$, it is necessary to investigate the more complicated setup of the most general detection operator as shown in Fig. 3.15. In this case, the detection operator for the ith detector D_i can be written in the form

$$\hat{D}_i = \alpha_i |+\rangle_{a_1}\langle e| + \beta_i |-\rangle_{a_1}\langle e| + \gamma_i |+\rangle_{a_2}\langle e| + \delta_i |-\rangle_{a_2}\langle e| \qquad (3.42)$$

with $\alpha_i, \beta_i, \gamma_i, \delta_i \in \mathbb{C}$. The detection of two scattered photons from the double excited state $|ee\rangle$ is described by applying two detection operators $\hat{D}_1$ and $\hat{D}_2$:

$$|\psi^{(f)}\rangle = \hat{D}_2\hat{D}_1|ee\rangle = \frac{1}{\zeta}[(\gamma_2\alpha_1 + \alpha_2\gamma_1)|++\rangle + (\delta_2\alpha_1 + \alpha_2\delta_1)|+-\rangle$$
$$+ (\gamma_2\beta_1 + \beta_2\gamma_1)|-+\rangle + (\delta_2\beta_1 + \beta_2\delta_1)|--\rangle] \quad (3.43)$$

with ζ a normalization factor. Thus, the anystate for $N = 2$ can be generated if and only if the equation

$$(\gamma_2\alpha_1 + \alpha_2\gamma_1)|++\rangle + (\delta_2\alpha_1 + \alpha_2\delta_1)|+-\rangle$$
$$+ (\gamma_2\beta_1 + \beta_2\gamma_1)|-+\rangle + (\delta_2\beta_1 + \beta_2\delta_1)|--\rangle =$$
$$a|++\rangle + b|+-\rangle + c|-+\rangle + d|--\rangle = |A_2\rangle \quad (3.44)$$

has a solution for arbitrary a, b, c, d. The normalization constant ζ is dropped since we assume the right hand side to be normalized ($|a|^2 + |b|^2 + |c|^2 + |d|^2 = 1$). Unfortunately, Eq. (3.44) is non-linear and therefore also non-trivial to solve. In what follows, we will search for a solution.

CHAPTER 3. QUANTUM STATE ENGINEERING

We start by rewriting the equation above as a system of quadratic equations:

$$\begin{aligned}\gamma_2\alpha_1 + \alpha_2\gamma_1 &= a & \delta_2\alpha_1 + \alpha_2\delta_1 &= b \\ \gamma_2\beta_1 + \beta_2\gamma_1 &= c & \delta_2\beta_1 + \beta_2\delta_1 &= d\end{aligned} \qquad (3.45)$$

The probably most elegant solution to this problem is to bring it into the form of a matrix equation[8]:

$$\begin{pmatrix} \gamma_2 & \gamma_1 \\ \delta_2 & \delta_1 \end{pmatrix} \begin{pmatrix} \alpha_1 & \beta_2 \\ \alpha_2 & \beta_1 \end{pmatrix} = \begin{pmatrix} a & c \\ b & d \end{pmatrix} \qquad (3.46)$$

One immediately sees that by choosing the parameters of one matrix as the unit matrix and the parameters of the other as a, b, c, d an arbitrary solution is possible. However, as there is no such simple solution for the three-qubit problem, we will present a different ansatz which can be applied for arbitrary N. First, we will transform Eq. (3.45) into a system of pseudo-linear equations by considering the prefactors of $\hat{D}_1$ as parameters and those of $\hat{D}_2$ as free variables: this leads to the matrix equation

$$\begin{pmatrix} \gamma_1 & 0 & \alpha_1 & 0 \\ \delta_1 & 0 & 0 & \alpha_1 \\ 0 & \gamma_1 & \beta_1 & 0 \\ 0 & \delta_1 & 0 & \beta_1 \end{pmatrix} \cdot \begin{pmatrix} \alpha_2 \\ \beta_2 \\ \gamma_2 \\ \delta_2 \end{pmatrix} = \begin{pmatrix} a \\ b \\ c \\ d \end{pmatrix}. \qquad (3.47)$$

Transforming the matrix in Eq. (3.47) into the row echelon form gives a system of equations of the form

$$\begin{pmatrix} \gamma_1 & 0 & \alpha_1 & 0 \\ 0 & \gamma_1 & \beta_1 & 0 \\ 0 & 0 & \frac{\delta_1}{\gamma_1}\alpha_1 & \alpha_1 \\ 0 & 0 & 0 & 0 \end{pmatrix} \cdot \begin{pmatrix} \alpha_2 \\ \beta_2 \\ \gamma_2 \\ \delta_2 \end{pmatrix} = \begin{pmatrix} a \\ c \\ b - \frac{\delta_1}{\gamma_1}a \\ d - \frac{\delta_1}{\gamma_1}c - \frac{\beta_1}{\alpha_1}b + \frac{\beta_1\delta_1}{\alpha_1\gamma_1}a \end{pmatrix}. \qquad (3.48)$$

Obviously, there exists only a solution to this system of equations, if the fourth entry of the vector on the right hand side is equal to zero. Solving for one parameter, this condition can be expressed as

$$\beta_1 = \frac{\delta_1 c - \gamma_1 d}{\delta_1 a - \gamma_1 b}\alpha_1 \quad \text{or} \quad \delta_1 = \frac{d}{c}\gamma_1 \quad \text{if} \quad a = b = 0 \wedge c \neq 0. \qquad (3.49)$$

In case $a = b = c = 0$, the state which is to be generated is simply $|--\rangle$ and the solution is trivial (e.g. all detector prefactors 0, except for $\delta_1 = \delta_2 = 1$). In all other

---
[8]Thanks to Thierry Bastin for this idea.

3.6. GENERATION OF ARBITRARY STATES: THE ANYSTATE?

cases, one can easily solve Eq. (3.48) and finds

$$\alpha_2 = \frac{a - \alpha_1\gamma_2}{\gamma_1}, \qquad \beta_2 = \frac{c - \beta_1\gamma_2}{\gamma_1}, \quad \text{and} \qquad \delta_2 = \frac{\gamma_1 b - \delta_1 a - \delta_1\alpha_1\gamma_2}{\alpha_1\gamma_1}, \qquad (3.50)$$

where γ_2 and α_1 arbitrary, γ_1 arbitrary but unequal 0, β_1 if applicable according to Eq. (3.49) otherwise arbitrary, δ_1 according to Eq. (3.49) if applicable otherwise arbitrary with the restriction that the denominator of Eq. (3.49) does not become zero.

3.6.4 Generating the Anystate for N=3

Let us now examine whether it is possible to extend this scheme to three qubits. As we now have to deal with more possible product states and one more detector, the number of parameters rises significantly and we expect the problem to become more involved. The most general detection operator for three atoms can be denoted by

$$\hat{D}_i = \alpha_i |+\rangle_{a_1}\langle e| + \beta_i |-\rangle_{a_1}\langle e| + \gamma_i |+\rangle_{a_2}\langle e| + \delta_i |-\rangle_{a_2}\langle e| + \epsilon_i |+\rangle_{a_3}\langle e| + \zeta_i |-\rangle_{a_3}\langle e|, \qquad (3.51)$$

with $\alpha_i, \beta_i, \gamma_i, \delta_i, \epsilon_i, \zeta_i \in \mathbb{C}$ and $i = 1, 2, 3$.

The most general quantum mechanical state for three atomic qubits is an arbitrary linear combination of $2^3 = 8$ basis states

$$|A_3\rangle = a|+++\rangle + b|++-\rangle + c|+-+\rangle + d|-++\rangle + e|--+\rangle + f|-+-\rangle + g|+--\rangle + h|---\rangle \qquad (3.52)$$

with a, b, c, d, e, f, g, h arbitrary complex numbers obeying the usual normalization.

After successful detection of three photons, the final state is found to be

$$|\psi^{(f)}\rangle = \hat{D}_3\hat{D}_2\hat{D}_1|eee\rangle =$$
$$(\alpha_1\gamma_2\epsilon_3 + \alpha_1\epsilon_2\gamma_3 + \gamma_1\alpha_2\epsilon_3 + \gamma_1\epsilon_2\alpha_3 + \epsilon_1\alpha_2\gamma_3 + \epsilon_1\gamma_2\alpha_3)|+++\rangle$$
$$+(\alpha_1\gamma_2\zeta_3 + \alpha_1\zeta_2\gamma_3 + \gamma_1\alpha_2\zeta_3 + \gamma_1\zeta_2\alpha_3 + \zeta_1\alpha_2\gamma_3 + \zeta_1\gamma_2\alpha_3)|++-\rangle$$
$$+(\alpha_1\epsilon_2\delta_3 + \alpha_1\delta_2\epsilon_3 + \epsilon_1\alpha_2\delta_3 + \epsilon_1\delta_2\alpha_3 + \delta_1\alpha_2\epsilon_3 + \delta_1\epsilon_2\alpha_3)|+-+\rangle$$
$$+(\gamma_1\epsilon_2\beta_3 + \gamma_1\beta_2\epsilon_3 + \epsilon_1\gamma_2\beta_3 + \epsilon_1\beta_2\gamma_3 + \beta_1\gamma_2\epsilon_3 + \beta_1\epsilon_2\gamma_3)|-++\rangle \qquad (3.53)$$
$$+(\alpha_1\delta_2\zeta_3 + \alpha_1\zeta_2\delta_3 + \delta_1\alpha_2\zeta_3 + \delta_1\zeta_2\alpha_3 + \zeta_1\alpha_2\delta_3 + \zeta_1\delta_2\alpha_3)|+--\rangle$$
$$+(\gamma_1\beta_2\zeta_3 + \gamma_1\zeta_2\beta_3 + \beta_1\gamma_2\zeta_3 + \beta_1\zeta_2\gamma_3 + \zeta_1\gamma_2\beta_3 + \zeta_1\beta_2\gamma_3)|-+-\rangle$$
$$+(\epsilon_1\beta_2\delta_3 + \epsilon_1\delta_2\beta_3 + \beta_1\epsilon_2\delta_3 + \beta_1\delta_2\epsilon_3 + \delta_1\epsilon_2\beta_3 + \delta_1\beta_2\epsilon_3)|--+\rangle$$
$$+(\beta_1\delta_2\zeta_3 + \beta_1\zeta_2\delta_3 + \delta_1\beta_2\zeta_3 + \delta_1\zeta_2\beta_3 + \zeta_1\beta_2\delta_3 + \zeta_1\delta_2\beta_3)|---\rangle.$$

The prefactors of the product states arise from six different quantum paths, as for a fixed product state there exist alltogether six different possibilities for the detection process: $\hat{D}_1$ detects one out of three photons, each being emitted from a different atom, leaving two possibilities for the second detector and one for the third detector totalling up to 3! possibilities. In analogy to the case of two atoms, we can form a system of pseudo-linear equations from the condition $|\psi^{(f)}\rangle = |A_3\rangle$ by extracting the variables describing D_3:

$$\begin{pmatrix} \{\epsilon,\gamma\} & \{\epsilon,\alpha\} & \{\gamma,\alpha\} & 0 & 0 & 0 \\ \{\zeta,\gamma\} & \{\zeta,\alpha\} & 0 & 0 & 0 & \{\gamma,\alpha\} \\ \{\delta,\epsilon\} & 0 & \{\delta,\alpha\} & 0 & \{\epsilon,\alpha\} & 0 \\ 0 & \{\beta,\epsilon\} & \{\beta,\gamma\} & \{\epsilon,\gamma\} & 0 & 0 \\ \{\zeta,\delta\} & 0 & 0 & 0 & \{\zeta,\alpha\} & \{\delta,\alpha\} \\ 0 & \{\zeta,\beta\} & 0 & \{\zeta,\gamma\} & 0 & \{\beta,\gamma\} \\ 0 & 0 & \{\delta,\beta\} & \{\delta,\epsilon\} & \{\beta,\epsilon\} & 0 \\ 0 & 0 & 0 & \{\zeta,\delta\} & \{\zeta,\beta\} & \{\delta,\beta\} \end{pmatrix} \begin{pmatrix} \alpha_3 \\ \gamma_3 \\ \epsilon_3 \\ \beta_3 \\ \delta_3 \\ \zeta_3 \end{pmatrix} = \begin{pmatrix} a \\ b \\ c \\ d \\ e \\ f \\ g \\ h \end{pmatrix}, \quad (3.54)$$

where $\{x,y\} = x_1 y_2 + y_1 x_2$. In a next step we transform the 8×6-matrix into row echelon form by application of a version of Gaussian elimination. The final system of equations then takes the form:

$$\begin{pmatrix} 1 & 0 & 0 & 0 & 0 & 0 \\ 0 & 1 & 0 & 0 & 0 & 0 \\ 0 & 0 & 1 & 0 & 0 & 0 \\ 0 & 0 & 0 & 1 & 0 & 0 \\ 0 & 0 & 0 & 0 & 1 & 0 \\ 0 & 0 & 0 & 0 & 0 & 1 \\ 0 & 0 & 0 & 0 & 0 & 0 \\ 0 & 0 & 0 & 0 & 0 & 0 \end{pmatrix} \begin{pmatrix} \alpha_3 \\ \gamma_3 \\ \epsilon_3 \\ \beta_3 \\ \delta_3 \\ \zeta_3 \end{pmatrix} = \begin{pmatrix} x_1 \\ x_2 \\ x_3 \\ x_4 \\ x_5 \\ x_6 \\ g - \{\delta,\beta\} x_3 - \{\delta,\epsilon\} x_4 - \{\beta,\epsilon\} x_5 \\ h - \{\zeta,\delta\} x_4 - \{\zeta,\beta\} x_5 - \{\delta,\beta\} x_6 \end{pmatrix} \quad (3.55)$$

with $x_{1,2,3,4,5,6}$ being terms that arise by applying the Gaussian elimination scheme. They contain the variables of the detection operators $\hat{D}_1$ and $\hat{D}_2$ and the variables a to h of the desired state. In order to arrive at a solution, the only thing that remains to be shown is that one can choose all variables from $\hat{D}_1$ and $\hat{D}_2$ such that the right hand side vector is well defined (no zero in any denominator) while at the same time its lowest two entries are zero. This rather tedious task is conducted in App. A [126].

3.6.5 Conclusion

While it is still possible to find a specific solution to the emerging system of non-linear equations by hand for $N = 3$, the task becomes quite hopeless for $N = 4$, in which case 32 free detection parameters have to be adjusted such that 16 non-linear equations are

3.6. GENERATION OF ARBITRARY STATES: THE ANYSTATE?

fulfilled. A sophisticated algorithmic approach is provided by the technique of Gröbner bases [127]. These algorithms are implemented in the PC-programm "Mathematica" and a solution has been attempted on a standard personal computer (3.2 GHz Dual Core, 4 GB RAM). Unfortunately, even though systems with up to 20 variables in 14 equations have been solved already in 1986 [128], Mathematica does not solve the present problem for $N = 4$ within 24 hours of calculation time. However, out of the dimensionality arguments given before, the way the prefactors in the equations are formed, and the fact that $2N^2$ variables always seem to provide a $2N^2$ dimensional solution (cf. cases $N = 2,3$), intuition leads to the conclusion that there should exist a solution for the anystate up to $N = 6$.

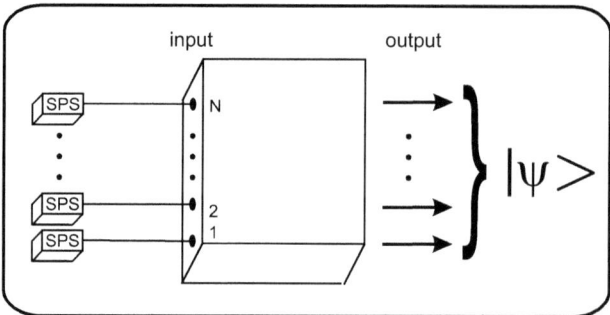

Figure 3.16: The proposed method provides a black-box which can entangle N uncorrelated single photons emitted from arbitrary single photon sources to a final state $|\psi\rangle$.

3.7 Interchanging the Roles of Atoms and Photons: A Versatile Source [12]

3.7.1 Introduction

In quantum optics the largest diversity of multipartite entangled states has been obtained so far by using the polarization degrees of freedom of photons, generated by spontaneous parametric down-conversion (SPDC) and subsequently fed into special arrangements of linear optical elements [7, 25–27, 29, 49, 66–76]. Although a large variety of different states can be produced by this technique, it usually suffers from the need of a particular experimental configuration for the generation of a particular entangled state [25–27, 29, 66–71, 73, 76]. Only recently, experiments have been performed capable of creating more than one state out of the same entanglement class [74], a family of states [49], or even different entanglement classes [7, 29, 75], inequivalent under stochastic local operations and classical communication (SLOCC) [96, 129, 130], within the same experimental setup. In the ideal case, however, one would like to have a single apparatus that tunes in *any wanted multipartite entangled state by simply turning a knob* [131].

Based on the results of the previous chapters, we now discuss in this chapter a technique, realizable with current technology, that allows to create an extremely large variety of multi-photon entangled states within the same experimental setup [12]. This can be achieved by simply turning the orientation of polarization filters and/or extending the length of optical pathways, without changing otherwise the experimental design. The technique allows to generate for any number of photonic qubits, amongst others, all

3.7. A VERSATILE SOURCE OF POLARIZATION-ENTANGLED PHOTONS

states symmetric under permutation of the photons (e.g. GHZ- and W-states), cluster states, and the entire class of – symmetric and non-symmetric – total angular momentum eigenstates of the multi-qubit compound. In addition, the scheme also enables to generate the canonical states representing all possible entanglement families of symmetric states inequivalent under SLOCC [12, 132], as recently defined in [133, 134]. The method does not rely on a specific source but admits to employ arbitrary single photon sources emitting photons of identical frequency. The technique represents in this way a black-box capable of generating a large variety of entangled states starting from initially uncorrelated photons (see Fig. 3.16).

3.7.2 Setup

The proposed scheme consists of N uncorrelated single photon sources emitting photons of identical frequency, e.g. trapped ions, neutral atoms, quantum dots, molecules or even photons produced via SPDC. If Λ-level atoms or similar multi-level emitters are used as single photon source, then in front of each source a polarization filter has to be installed which projects the polarization vector of the emitted photon onto the polarizer's axis ε. If the source emits photons of well defined polarization it suffices to turn the photon polarization vector along the desired axis ε by use of a quarter- and a half-wave plate; the latter has the advantage that the count rate can be increased because no photons are lost in a filtering process. Subsequently, the photons are registered by N detectors located in the far-field region of the sources at $\mathbf{r}_1, \ldots, \mathbf{r}_N$ picking out N spatial modes defined by the wave vectors $\mathbf{k}_n$, $n \in 1, \ldots, N$ which point from the atoms towards the detectors at $\mathbf{r}_1, \ldots, \mathbf{r}_N$. As before, the far-field condition ensures that the optical phase does not vary appreciably over the surface of the detector.

On their way from the sources to the detectors each photon will accumulate an optical phase ϕ_{nm} given by

$$\phi_{nm} = kR_{nm}, \qquad (3.56)$$

where R_{nm} is the optical path length from source n to detector m ($n, m \in 1, \ldots, N$) and k is the wave number of the photons. In case of using optical fibers guiding the photons from the sources to the detectors (see Fig. 3.17), R_{nm} corresponds to the optical path length through the fiber. By assuming that each detector registers exactly one photon, the correlated photon detection signal will display N-photon interferences. By changing the orientation of the polarization filters/polarization turning device (called in the following *polarization device*) and/or the optical phases ϕ_{nm}, a large variety of polarization-entangled photonic N-qubit states can be produced.

To see this in more detail let us start by considering a particular photon emitted by source n. Its wavefunction compatible with a successful measurement at any of the

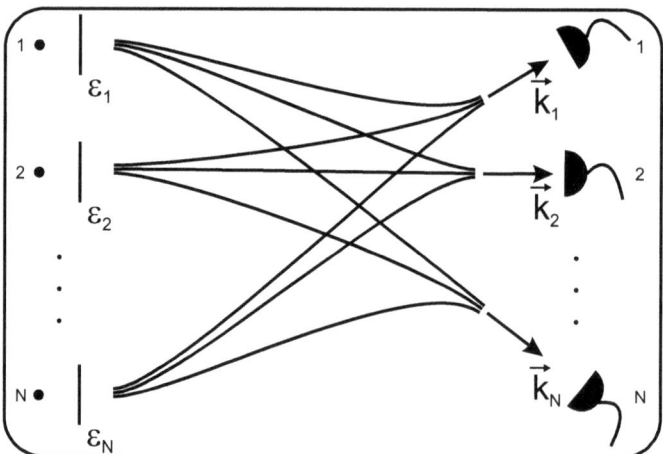

Figure 3.17: The proposed setup implemented with optical monomode fibers. The black dots on the left hand side represent N single photon sources each with a polarization device in front oriented along $\boldsymbol{\varepsilon}$. The optical single mode fibers lead from each source to each detector.

N detectors is given by

$$|\psi\rangle = \hat{\tilde{P}}_n |0,0,\ldots,0\rangle \,, \tag{3.57}$$

where

$$\hat{\tilde{P}}_n = \sum_{m=1}^{N} e^{i\phi_{nm}} \left(\alpha_n \hat{a}_{\sigma^+}^{\dagger(m)} + \beta_m \hat{a}_{\sigma^-}^{\dagger(m)} \right). \tag{3.58}$$

Hereby, the state vector $|\psi\rangle = |x_1, x_2, \ldots, x_N\rangle = |x_1\rangle_1 \otimes |x_2\rangle_2 \otimes \ldots \otimes |x_N\rangle_N$ describes the photon polarization state $|x_m\rangle_m = \alpha_m |\sigma^+\rangle_m + \beta_m |\sigma^-\rangle_m$ in the mode $\mathbf{k}_m$ for $m \in 1, \ldots, N$, $\hat{a}_{\sigma^\pm}^{\dagger(m)}$ is the creation operator of a photon with polarization $\sigma^\pm$ in mode $\mathbf{k}_m$, $|0,0,\ldots,0\rangle$ is the vacuum state and the sum runs over all N detectors as there is an equal probability for the photon to be registered by anyone of the N detectors. The polarization vector of the photon after passing the polarization device in front of source n is given by $\boldsymbol{\varepsilon}_\mathbf{n} = \alpha_n \boldsymbol{\sigma}^+ + \beta_n \boldsymbol{\sigma}^-$, with arbitrary complex coefficients α_n and β_n ($|\alpha_n|^2 + |\beta_n|^2 = 1$); the coefficient α_n (β_n) defines thus the probability amplitude of detecting a photon from source n with polarization σ^+ (σ^-).

To obtain the state of the system compatible with a successful detection event of N photons we have to apply the N operators $\hat{\tilde{P}}_n$, $n \in 1, \ldots, N$, onto the vacuum state

$|0, 0, \ldots, 0\rangle$, representing the emission of N photons by the N sources. However, since we consider only the case where each detector registers exactly one photon we have to change the operator $\hat{\tilde{P}}_n$ to $\hat{P}_n$ given by

$$\hat{P}_n = \sum_{m=1}^{N} e^{i\phi_{nm}} \left(\alpha_n |\sigma^+\rangle_m \langle 0| + \beta_n |\sigma^-\rangle_m \langle 0| \right), \tag{3.59}$$

where the operator $|\sigma^\pm\rangle_m \langle 0|$ creates a photon with polarization $\sigma^\pm$ in mode $\mathbf{k}_m$ if and only if the mode $\mathbf{k}_m$ was unpopulated before. In this way we ensure that each mode $\mathbf{k}_m$ is populated with no more than one photon.

The entire N photon state compatible with a measurement of one photon in each mode is then given by

$$|\psi_f\rangle = \hat{P}_N \ldots \hat{P}_2 \hat{P}_1 |0, 0, ..., 0\rangle. \tag{3.60}$$

3.7.3 Generation of Photonic Quantum States

A closer inspection of the equalities given above reveals striking similarities to the mathematical formalism discussed in the previous sections: Eq. (3.59) has exactly the same structure as the detection operator from Eq. (3.7) describing the projection of a Λ-level emitter upon detection of a photon. By comparing Eq. (3.7) with Eq. (3.59), we can see that the mth atom accords to the mth mode in Fig. 3.17, an atom in the excited state $|e\rangle$ to an empty mode $|0\rangle$, and the two atomic ground states $|\pm\rangle$ to the two photonic polarization states $|\sigma^\mp\rangle$. Moreover, as the N-fold application of the detection operator from Eq. (3.7) to the completely excited state $|ee \ldots e\rangle$ is formally equivalent to Eq. (3.60), we can generate the same quantum states among the polarization degrees of freedom of N photons as we can generate among the ground states of N Λ-level atoms.

Generation of Symmetric States

Let us start by assuming that all optical phases ϕ_{nm} are multiples of 2π, an assumption which is particularly useful for a far-field implementation. Then, the operators $\hat{P}_n$, $n \in 1, \ldots N$, and therefore also the final photonic state $|\psi_f\rangle$, are totally symmetric under permutation of the modes. In this case $|\psi_f\rangle$ can be expressed as a linear combination of the $N+1$ symmetric Dicke states with k σ^- excitations $|D_N(k)\rangle$ [99]:

$$|\psi_f\rangle = \mathcal{N} \sum_{k=0}^{N} c_k |D_N(k)\rangle, \tag{3.61}$$

where $\mathcal{N}$ is a normalization prefactor. As it turns out from Eq. (3.59) and Eq. (3.60), the c_k are given by

$$c_k = \binom{N}{k}^{\frac{1}{2}} \sum_{1 \leq i_1 \neq \ldots \neq i_N \leq N} \beta_{i_1} \ldots \beta_{i_k} \alpha_{i_{k+1}} \ldots \alpha_{i_N}. \qquad (3.62)$$

Thus, to generate a particular symmetric Dicke state $|D_N(K)\rangle$, $K \in 1, \ldots, N$, we have to choose the α_n and β_n such that only $c_K \neq 0$, while all other $c_k = 0$. This can be achieved by setting $\beta_1 = \ldots = \beta_K = 1$ and $\beta_{K+1} = \ldots = \beta_N = 0$ (and therefore $\alpha_1 = \ldots = \alpha_K = 0$ and $\alpha_{K+1} = \ldots = \alpha_N = 1$), i.e, by choosing the orientation of the polarization devices such that K photons have polarization σ^- and $(N-K)$ photons polarization σ^+ [81]. The symmetry of the setup then leads to a symmetric distribution of the photons among the modes.

Following [82], this technique also allows to generate *any* symmetric state $|\psi_{sym}\rangle = \sum_{k=0}^{N} d_k |D_N(k)\rangle$ with respect to permutations of the modes. For this, the polarization devices must be oriented such that for the polynomial of degree K

$$P(z) = \sum_{k=0}^{N} (-1)^{K-k} \sqrt{\binom{N}{k} / \binom{N}{K}} d_k z^k, \qquad (3.63)$$

the α_n/β_n identify to the K roots of $P(z)$ for $n \leq K$, while the remaining β_n (α_n) are set to 0 (1).

Generation of Total Angular Momentum Eigenstates

For the creation of non-symmetric states, we have to make full use of the degrees of freedom of the system, e.g., by varying the optical phases ϕ_{nm} to values unequal to 2π (e.g. by modifying the lengths of the optical fibers) or even removing optical fibers, i.e., not connecting all sources to all detectors.

In particular, if we allow to vary the optical phases ϕ_{nm} to any value between 0 and 2π and/or to remove single optical fibers, suppressing thereby certain quantum paths, it becomes possible to generate any – symmetric and non-symmetric – total angular momentum eigenstates of the N photonic qubit compound, as demonstrated in case of atomic qubits in Sec. 3.4.

Using the notation of Sec. 3.4.2 where the total angular momentum eigenstates are denoted by $|S_1, S_2, S_3, \ldots, S_N; m_s\rangle$ (where $S_1, S_2, \ldots, S_{N-1}$ takes the coupling history into account and S_N and m_N define the eigenvalues of the square of the total spin operator $\hat{S}^2$ and its z-component $\hat{S}_z$, $S_N(S_N+1)\hbar^2$ and $m_N\hbar$, respectively), we can formulate a protocol to generate any of the photonic N-qubit total angular momentum eigenstates by adapting the algorithm introduced in Sec. 3.4.4. In order to produce the

3.7. A VERSATILE SOURCE OF POLARIZATION-ENTANGLED PHOTONS

state $|S_1, S_2, S_3, \ldots, S_N; m_s\rangle$ we have to

1. set up in front of the N sources $\frac{N}{2} + m_s$ ($\frac{N}{2} - m_s$) σ^-- (σ^+-) oriented polarization devices. Hereby, we connect all sources with optical fibers to the first detector.

2. check for each detector j beginning with $j = 2$ whether $S_j > S_{j-1}$ or $S_j < S_{j-1}$. If

   a. $S_j > S_{j-1}$, we have to connect detector j with optical fibers to all sources except those which are mentioned in case [b.] below.

   b. $S_j < S_{j-1}$, we have to connect detector j with optical fibers to one source with a σ^- polarization device and to one with a σ^+ polarization device in front. The optical fiber leading to the σ^- polarization device should induce a relative optical phase shift of π and those two sources should not be linked to any other subsequent detectors.

In this way, we can form any of the 2^N symmetric and non-symmetric total angular momentum eigenstates of the photonic N-qubit compound.

Generation of Cluster States

In Sec. 3.5, we have seen how to create a cluster state in $(N+M)$ atomic qubits. Now, we know how to transfer this setup to generate a cluster state in $N+M$ photonic qubits: We can use the same setup as depicted in Fig. 3.14, only that the polarization filters need to be switched from their place between the fibers and the detectors to the place between the single photon emitters and the detectors. A detection of one photon at each output port will then witness the creation of a state of the form

$$|\psi\rangle = |H\rangle^{M+N} + |H\rangle^M |V\rangle^N + |V\rangle^M |H\rangle^N - |V\rangle^{M+N}. \tag{3.64}$$

Generation of the Most Arbitrary State

We have seen that with the method proposed here, it is possible to create many families and classes of photonic states. Naturally, the question about the limits arises. Because of the formal equivalence of Eq. (3.7) and Eq. (3.59), the same considerations given in Sec. 3.6 remain valid: while we know how to generate any arbitrary state for two and three photons, and we expect this to be possible for up to six photons, one can tell by simple dimensional arguments that for 7 and more photons, there are states that cannot be created with this detection technique.

3.7.4 Conclusion

As SPDC is the brightest single photon source available it is favorable to employ it as a photon source for our proposed method. However, as mentioned, our method entan-

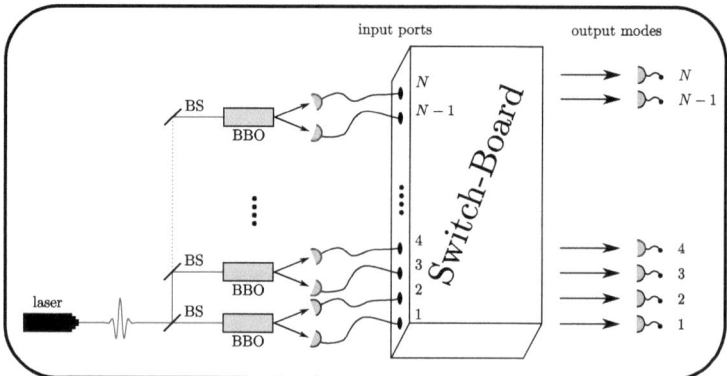

Figure 3.18: Photonic multiqubit states generated by use of type-I SPDC as single photon sources and an appropriate configuration of optical fibers, considered as an optical switch-board.

gles N *uncorrelated* single photons. Therefore, in order to eliminate the momentum correlations among the photons generated via non-collinear SPDC, we can couple the photons into optical fibers. As each non-linear crystal scatters two SPDC photons, which can both be used, we need to set up $N/2$ non-linear crystals to supply N single photons. Our method provides in this way an *optical switch-board* which brings the initially uncorrelated photons into a large variety of entangled states (see Fig. 3.18). For the generation of a 4 photon entangled state using two BBO crystals pumped by a pulsed Ti:Sapphire laser at 80 MHz repetition rate, count rates of more than 4 Hz can be expected [71].

In conclusion we proposed a scheme which allows to generate all symmetric states, cluster states, as well as all (symmetric and non-symmetric) total angular momentum eigenstates among N photonic qubits within the same experimental setup. In particular, the scheme allows to generate all canonical states representing the possible entanglement families of symmetric states, inequivalent under SLOCC [133, 134]. To tune from one entanglement family of symmetric states to another, one just has to turn the orientation of the polarization devices in front of the sources. In this way, by appropriately configuring the optical fibers which connect the sources with the detectors, this optical switch-board allows to produce an extremely large variety of photonic multi-qubit states. We note that the protocol can be implemented by use of many different single photon sources, including photons generated by non-collinear SPDC. The latter has the advantage to obtain a significantly higher count rate compared to other single photon sources currently available.

Chapter 4

Complementary Measurements: Wave-Particle Duality

4.1 Introduction

In the last chapter, we have invoked the concept of indistinguishable photon paths in order to explain the emergence of quantum interferences and the corresponding generation of entangled states among the emitters without attempting to determine whether the photons did actually take a well defined path. This might have well been possible, as the polarization of the photons has been assumed to be entangled with the internal atomic state (cf. Sec. 3.2.2) so that the atoms might have acted as a which-way detector (WWD). To visualize this concept, assume that two photons from two Λ-level atoms are measured, one with a σ^+-polarization, the other one with a σ^--polarization. In this case, after the measurement, it is possible to determine which quantum path was realized (i.e., which atom emitted which photon) by reading out the internal states of the atoms, since we know that an atom will be found in the state $|+\rangle$ ($|-\rangle$) if it has emitted a σ^-- (σ^+-) polarized photon. This simple example already raises a number of fundamental questions: Do we still observe an interference pattern even though we can determine the path of the photon completely? If not, why do we still talk about quantum interferences? What happens, if we are able to determine the quantum path only partially, e.g., by registering two photons from Λ-level systems, one in the state $|\sigma^-\rangle$ and the other in the state $\alpha|\sigma^-\rangle + \beta|\sigma^+\rangle$? And if the atoms are in a *coherent* superposition $|+-\rangle + |-+\rangle$ after detection of the photons, is it even sensible to talk about which-way (WW) information if we project the atoms onto one of those two states in a measurement?

These questions lead the way into the last part of this thesis dealing with the much debated field of complementary measurements. The discussion about the interrelationship between knowing the path of a photon in an interferometer and observing its

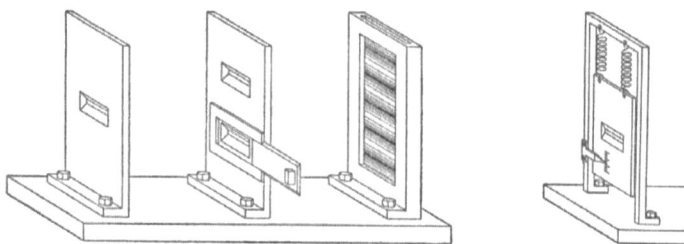

Figure 4.1: Recoiling slit as proposed by Einstein. On the left a simple Young-type interferometer is shown. A single collimating slit provides spatial coherence. Thereafter, each photon is diffracted at a double slit and the interference pattern is observed on a screen. By closing one of the slits, one can make sure that one knows through which slit the photon travels; at the same time the interference pattern disappears. Einstein proposed to replace the single (fixed) slit by a slit suspended on extremely sensitive springs as shown on the right. If a photon is diffracted at this slit, it transfers some momentum onto the slit which depends on whether it is diffracted towards the upper or the lower path and which can in principal be measured. However, Bohr could show that Heisenberg's uncertainty relation prevents simultaneous knowledge of the initial position and the initial momentum of the recoiling slit to the degree of accuracy necessary to observe an interference pattern and the path of the photon concurrently. (Drawings from [15])

interference pattern started with the famous debate between Niels Bohr and Albert Einstein which triggered the Copenhagen interpretation of quantum mechanics [15, 135]. In these discussions, Einstein tried to construct a device which was capable of measuring the path of a photon in a Young-type double slit interferometer [136] while at the same time observing an interference pattern. Thereby, he hoped to show that the quantum mechanical description of Nature was incomplete. The most famous of these attempts is his introduction of a recoiling slit in the Young-type interferometer which was supposed to measure the momentum transfer of the photon to the slit (cf. Fig. 4.1).

However, for all devices proposed by Einstein, Bohr succeeded in finding an uncertainty relation of the Heisenberg-Robertson kind (cf. Eq. (1.5)) which prevented either a successful path measurement or the observation of an interference pattern [15]. Bohr concluded that a quantum system can have properties that are equally real and yet mutually exclusive, with the wave and particle properties of a system being but the most famous example. In a more modern wording, the duality in an interferometer can be described as the fact that "the acquisition of which-way information and the observation of an interference pattern are mutually exclusive" [18].

However, this statement only addresses the extremal cases of full WW information and no interference pattern and vice versa. Some thirty years ago, Wootters and Zurek published a first treatise in which they investigated the intermediate cases of acquiring

4.2. BASIC CONCEPTS OF QUANTITATIVE WAVE-PARTICLE DUALITY

partial WW information and observing a non-perfect interference pattern for the case of the recoiling slit [16]. Since then, WW detection schemes [18, 19, 137–147] as well as the strongly related quantum erasure schemes [137, 148–153] have received much attention. In both cases the setup consists of an interferometer featuring an auxiliary quantum system, the WWD, which detects the path of the interfering quantum mechanical object. If the experimenter strives to measure an observable of the WWD whose eigenvalues can be correlated to the path of that object, he implements a WW detection scheme while in case he chooses to measure eigenvalues of an observable which always projects the WWD into a state that renders the acquisition of WW information impossible a quantum eraser has been realized.

In the following, we present a scheme which can be considered a "mixture" of a WW detection scheme and a quantum eraser. For this purpose we introduce a new observable to be read out from the WWD whose eigenstates either carry full WW information or none at all. We show that this observable has the interesting property that the WW information it carries is correlated with the phase shift that the interferometer induces between the two quantum paths. In particular, it is demonstrated that with the interfering object being detected in a certain region around the minima of the interference pattern the new observable provides *more* WW information than the observable recognized in [18] to be optimal, while in the maxima it carries less [20]. We also show that for a fixed value of the visibility of the interference pattern this effect can be exploited to extract more WW information than previously thought possible [21].

4.2 Basic Concepts of Quantitative Wave-Particle Duality

In order to display quantitative relations between the wave properties and the particle properties of the interfering quantum mechanical object, called in the following a quanton[1], it is first necessary to quantify both the "strength" of the interference pattern which is observed at the end of the interferometer and our knowledge about the path the quanton took when passing through the interferometer.

4.2.1 Quantitative Interference

In order to quantify the "strength" of the interference in a periodic interference pattern, it is a traditional method to compare the maximal intensity $I_{\max}$ and the minimal intensity $I_{\min}$ obtained on varying the interferometric phase φ. This approach works particularly well in case of sinusoidal interference patterns, which will be the only ones we encounter in the course of this chapter.

---
[1] In order to avoid the terms wave and particle and the (classical) mental pictures associated with it, we will follow the nomenclature proposed by Lévy-Leblond [154] and Englert [18] and call the interfering quantum mechanical object a quanton.

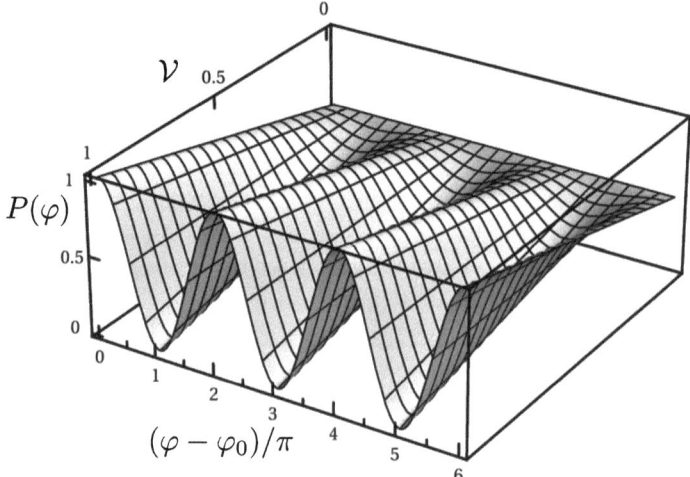

Figure 4.2: The interference pattern in dependence on the visibility $\mathcal{V}$ and the interferometric phase φ.

There exist two well-established ways of quantifying the relation between $I_{\max}$ and $I_{\min}$. One is the calculation of the contrast, given by $\frac{I_{\max}-I_{\min}}{I_{\max}}$. However, it turns out that often quantitative relations take on a particularly simple form when the visibility $\mathcal{V}$ of the interference fringes is employed, defined by

$$\mathcal{V} = \frac{I_{\max} - I_{\min}}{I_{\max} + I_{\min}}. \tag{4.1}$$

Using this measure, a sinusoidal interference pattern is always proportional to

$$P(\varphi) = \frac{1}{2}\left(1 + \mathcal{V}\cos(\varphi - \varphi_0)\right), \tag{4.2}$$

where φ_0 is the phase offset of the interference pattern. A plot of this interference pattern is shown in Fig. 4.2.

4.2.2 Quantitative Which-Way Knowledge

While there exist well-established procedures for quantifying the "strength" of an interference pattern, this is less the case for the quantification of WW knowledge. Several methods have been proposed, among them the Shannon entropy already used by Wootters and Zurek [16] and the degree of intrinsic indistinguishability introduced by Mandel [138]. Here, however, we will resort to the which-way knowledge (WW knowl-

4.2. BASIC CONCEPTS OF QUANTITATIVE WAVE-PARTICLE DUALITY 77

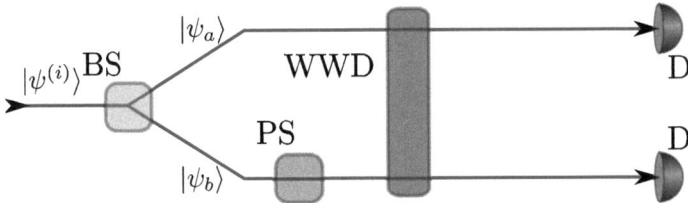

Figure 4.3: Scheme of gaining WW information with a WWD: A beam splitter (BS) takes a quanton into an equally weighted superposition of two paths and a phase shifter (PS) introduces a relative phase between them. Thereafter, a WWD interacts with the quanton. The information obtained by reading out the WWD can then be used to guess the path of the quanton. Afterwards, one can check whether the guess is correct by detecting the quanton on either of the two paths with the detectors D. In this way, we experimentally calibrate how much WW information can be gained from a certain detector read out result. After this callibration, we know how likely we are to know the path of the quanton correctly and can allow for the two paths to interfere instead of checking for the correctness of our guess.

edge) $\mathcal{K}_W$ introduced by Englert and Bergou [153]. (The subscript will be explained below.) In a physical picture, it can be described in the following way: If we were to make an educated guess on which path we will find the quanton, taking into account all information we have from the initial state of the system and the WWD, then $\mathcal{K}_W$ describes the percentage by which the number of right guesses exceeds the number of wrong guesses.

The chance of guessing the way correctly can be increased over a mere random choice in two ways. Either we are dealing with an unbalanced interferometer in which the quanton prefers one path over the other by a certain degree. This preference has been termed the predictability $\mathcal{P}$ [18]. However, in the present work we will restrict ourselves to balanced interferometers in which the initial probabilities for the quanton to take either path are equal, i.e., the case $\mathcal{P} = 0$.

Instead, as already proposed by Einstein, one may introduce a WWD into the path of the quanton in order to determine the path it took (cf. Fig. 4.3). In the optimal case, this WWD performs a quantum non-demolition measurement in which, without any backaction on the quanton, the WWD is transferred from its initial state $|\chi^{(i)}\rangle$ into (in general) non-orthogonal states $|\chi_a\rangle$ and $|\chi_b\rangle$ which mark the two paths $|\psi_a\rangle$ and $|\psi_b\rangle$ of the quanton as

$$\frac{1}{\sqrt{2}}(|\psi_a\rangle + |\psi_b\rangle) \otimes |\chi^{(i)}\rangle \stackrel{\text{WWD}}{\to} \frac{1}{\sqrt{2}}(|\psi_a\rangle|\chi_a\rangle + |\psi_b\rangle|\chi_b\rangle) = |\Psi^{(f)}\rangle. \tag{4.3}$$

The state of the quanton and the state of the WWD after the interaction are calculated by a partial trace over the degrees of freedom of the WWD and the quanton respectively

$$\rho_Q = \text{tr}_{\text{WWD}}(|\Psi^{(f)}\rangle\langle\Psi^{(f)}|)$$
$$= \frac{1}{2}\left(|\psi_a\rangle\langle\psi_a| + |\psi_b\rangle\langle\psi_b| + \langle\chi_b|\chi_a\rangle|\psi_a\rangle\langle\psi_b| + \langle\chi_a|\chi_b\rangle|\psi_b\rangle\langle\psi_a|\right) \quad (4.4a)$$
$$\rho_D = \text{tr}_Q(|\Psi^{(f)}\rangle\langle\Psi^{(f)}|) = \frac{1}{2}\left(|\chi_a\rangle\langle\chi_a| + |\chi_b\rangle\langle\chi_b|\right). \quad (4.4b)$$

If $|\chi_a\rangle$ and $|\chi_b\rangle$ are orthogonal, then reading out an observable of the WWD with these eigenvectors will give full WW information. However, one might as well (among a myriad of other possibilities) choose to read out an observable with the eigenvectors $\frac{1}{\sqrt{2}}(|\chi_a\rangle \pm |\chi_b\rangle)$. These vectors contain no WW information at all and can serve to realize a quantum eraser [137]. Thus, obviously the WW knowledge that is extracted from the WWD *depends on the observable* $\hat{W}$ *chosen to read out the WWD*. This is the reason why $\mathcal{K}_W$ carries a subscript: it denotes the WW information which is extracted *for that specific observable* $\hat{W}$.

Furthermore, to include cases in which the two paths principally cannot be distinguished perfectly, we allow for arbitrary $|\chi_a\rangle$ and $|\chi_b\rangle$ which need not be orthogonal and consequently do not necessarily represent two different eigenstates of any observable.

All possible cases are encompassed in a description of the measurement in which we select a general observable $\hat{W}$ with the eigenbasis $W = \{|W_k\rangle\}$ ($k = 1, \ldots, n$) to be read out from the WWD. The subsequent readout of $\hat{W}$ will then return an eigenstate $|W_k\rangle$ which provides information about the path of the quanton (cf. Fig. 4.3); this information will typically be incomplete.

To quantify the completeness of the WW information we will make use of the WW knowledge $\mathcal{K}_W$. In order to derive $\mathcal{K}_W$, we start by calculating the likelihood $\mathcal{L}_W$ of guessing the way of the quanton correctly in an arbitrary run of the experiment (cf. Fig. 4.3). We note that $\mathcal{L}_W$ is determined by two probabilities: first we need to read out the observable $\hat{W}$ of the WWD and find it with a certain probability p_i in any of its eigenstates $|W_i\rangle$. The second probability describes the conditional probability q_i of guessing the way correctly with the detector being found in this particular eigenstate $|W_i\rangle$.

The probability p_i of finding a certain eigenstate of $\hat{W}$ is determined by

$$p_i = \langle W_i|\rho_D|W_i\rangle, \quad (4.5)$$

where the density matrix ρ_D denotes the state of the WWD after passage of the quanton (cf. Eq. (4.4b)). Assuming that such a measurement is realized and the WWD is projected into a certain eigenstate $|W_i\rangle$, one can look at the relative contributions of the two possible pathways to this final state and choose the one which contributes more [18]. Thus, we can express the probability of guessing the way correctly from

4.2. BASIC CONCEPTS OF QUANTITATIVE WAVE-PARTICLE DUALITY   79

$|W_i\rangle$ as

$$q_i = \frac{\max\left(|\langle\chi_a|W_i\rangle|^2, |\langle\chi_b|W_i\rangle|^2\right)}{|\langle\chi_a|W_i\rangle|^2 + |\langle\chi_b|W_i\rangle|^2}. \quad (4.6)$$

$\mathcal{L}_W$ is then calculated by combining these factors as [18]

$$\mathcal{L}_W = \sum_i p_i q_i. \quad (4.7)$$

$\mathcal{L}_W$ is a measure for the amount of WW information that is *extracted* from the WWD by measuring $\hat{W}$: If $\mathcal{L}_W = 0.5$ ($\mathcal{L}_W = 1$), we have to make a random guess about the path of the quanton (know the way for sure). In order to have a measure which has a value of 0 (1) for no (full) WW information, one can then introduce the WW knowledge $\mathcal{K}_W$ which rescales $\mathcal{L}_W$ to the interval from 0 to 1:

$$\mathcal{K}_W = 2\sum_i p_i q_i - 1. \quad (4.8)$$

$\mathcal{K}_W$ has exactly the statistical properties described at the beginning of this section.

Maximizing $\mathcal{K}_W$ with respect to W leads to a certain optimal WW knowledge $\mathcal{K}_{\text{opt}}$ which quantifies the amount of WW information that is principally *available* by reading out a *single* observable.[2] The central problem of this approach is the optimization involved in finding $\mathcal{K}_{\text{opt}}$ and it was solved in [18]. There, it was shown that $\mathcal{K}_W$ is maximized if an observable $\hat{E}$ of the WWD is read out whose eigenbasis consists of the eigenvectors of the operator

$$\hat{O} = ||\chi_a\rangle\langle\chi_a| - |\chi_b\rangle\langle\chi_b||. \quad (4.9)$$

In the following, we will call $\hat{E}$ the canonical observable and its eigenbasis E the canonical basis. In [18], it was also shown that the duality relation

$$\mathcal{K}_W^2 + \mathcal{V}^2 \leq 1 \quad (4.10)$$

always holds and is fulfiled optimally, if the WWD is initially in a pure state and the canonical observable $\hat{E}$ is read out.

4.2.3 Interferometer Formalisms

Before we apply the concepts described above, let us introduce the notational conventions for the interferometers we will use. Two specific kinds of interferometers will be of

---

[2]$\mathcal{K}_{\text{opt}}$ is often called the distinguishability $\mathcal{D}$. However, for clarity, we will not use this terminology here.

importance to us: The Mach-Zehnder-type interferometer (cf. Fig. 4.4), since it allows for a very general description, and the Young-type interferometer (cf. Fig. 4.5), as our examples will be based on it.

The Mach-Zehnder-type Interferometers

In general, all symmetric two-way interferometers can be described as follows: the quanton in the initial state $|\psi^{(i)}\rangle$ enters the interferometer and encounters a 50:50 beam splitter (BS) where its wavefunction is split equally into two orthogonal parts $|\psi_a\rangle$ and $|\psi_b\rangle$. The splitting may refer to an actual spatial seperation of the wavefunction as well as to different internal states of the quanton. Following Englert's notation in [18], we describe the transformation at the beam splitter by a Hadamard gate H acting on the incoming state $|\psi^{(i)}\rangle$

$$|\psi^{(i)}\rangle \stackrel{\text{BS}}{\to} H|\psi^{(i)}\rangle = \frac{1}{\sqrt{2}}\left(|\psi_a\rangle + |\psi_b\rangle\right), \qquad (4.11)$$

leading to the usual coherent superposition of both paths (cf. Eq. (4.3)). In the central stage of the interferometer, one part of the wavefunction experiences a tunable relative phase shift φ with respect to the other and the WWD changes its state as a result of the presence of the quanton.[3] Thus, the phase shifter transforms the state of the quanton according to

$$\frac{1}{\sqrt{2}}\left(|\psi_a\rangle + |\psi_b\rangle\right) \stackrel{\text{PS}}{\to} \frac{1}{\sqrt{2}}\left(|\psi_a\rangle + e^{i\varphi}|\psi_b\rangle\right), \qquad (4.12)$$

and, introducing the WWD as in Eq. (4.3), the state of the combined system of WWD and quanton is given by

$$\left(|\psi_a\rangle + e^{i\varphi}|\psi_b\rangle\right) \otimes |\chi^{(i)}\rangle \stackrel{\text{WWD}}{\to} |\psi_a\rangle|\chi_a\rangle + e^{i\varphi}|\psi_b\rangle|\chi_b\rangle. \qquad (4.13)$$

The two paths are then recombined into the output states $|\psi^+\rangle$ and $|\psi^-\rangle$ of the beam merger (BM), an action which is again described by a Hadamard gate, transforming the combined system into its final state $|\Psi^{(f)}\rangle$:

$$\frac{1}{\sqrt{2}}\left(|\psi_a\rangle|\chi_a\rangle + e^{i\varphi}|\psi_b\rangle|\chi_b\rangle\right) \stackrel{\text{BM}}{\to}$$

$$|\Psi^{(f)}\rangle = \frac{1}{2}\left(|\psi^+\rangle|\chi_a\rangle + |\psi^-\rangle|\chi_a\rangle + e^{i\varphi}|\psi^+\rangle|\chi_b\rangle - e^{i\varphi}|\psi^-\rangle|\chi_b\rangle\right) \quad (4.14)$$

$$= \frac{1}{2}\left(|\psi^+\rangle(|\chi_a\rangle + e^{i\varphi}|\chi_b\rangle) + |\psi^-\rangle(|\chi_a\rangle - e^{i\varphi}|\chi_b\rangle)\right).$$

---

[3]These two actions commute and consequently their order is of no importance.

4.2. BASIC CONCEPTS OF QUANTITATIVE WAVE-PARTICLE DUALITY 81

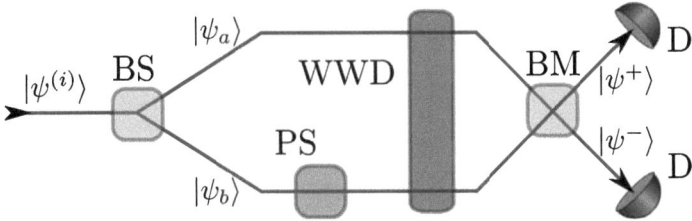

Figure 4.4: Scheme of a Mach-Zehnder-type interferometer, in which a WWD is inserted in order to measure the path of the quanton and the signal of the quanton is registered behind a beam merger (BM).

Finally, the interference pattern is revealed at the detector in the relative frequency $P^\pm(\varphi)$ of finding the quanton in either state $|\psi^+\rangle$ or $|\psi^-\rangle$ on varying the interferometric phase φ as

$$P_\varphi(|\psi^\pm\rangle) = \frac{1}{2}\left(1 \pm \mathcal{V}\cos(\varphi - \varphi_0)\right), \tag{4.15}$$

where the visibility of the interference pattern $P_\varphi(|\psi^\pm\rangle)$, defined in Eq. (4.1), is now given by

$$\mathcal{V} = |\langle \chi_a | \chi_b \rangle|, \tag{4.16}$$

and the offset φ_0 is defined by the phase of the overlap

$$\varphi_0 = \frac{\langle \chi_a | \chi_b \rangle}{|\langle \chi_a | \chi_b \rangle|}. \tag{4.17}$$

The Young-type Interferometer

In a general symmetric Young-type interferometer, a plane wave specifying the initial state of a quanton is seperated at two narrow slits into the two coherent parts $|\psi_a\rangle$ and $|\psi_b\rangle$ which describe the quanton emerging from the respective slit and propagating towards a screen in the far field where the quanton is registered. On the propagation, a relative phase $\varphi := \varphi(\mathbf{r})$ is accumulated between $|\psi_a\rangle$ and $|\psi_b\rangle$ which depends on the position $\mathbf{r}$ at the screen where the quanton is registered. The probability of detecting a quanton at a position $\mathbf{r}$ corresponding to a certain value of φ can be calculated as the overlap of the initial wavefunction $|\psi^{(i)}\rangle = \frac{1}{\sqrt{2}}(|\psi_a\rangle + |\psi_b\rangle)$ with its phase shifted version $|\psi_\varphi\rangle = \frac{1}{\sqrt{2}}(|\psi_a\rangle + e^{i\varphi}|\psi_b\rangle)$. In case a WWD is introduced, the wavefunction of

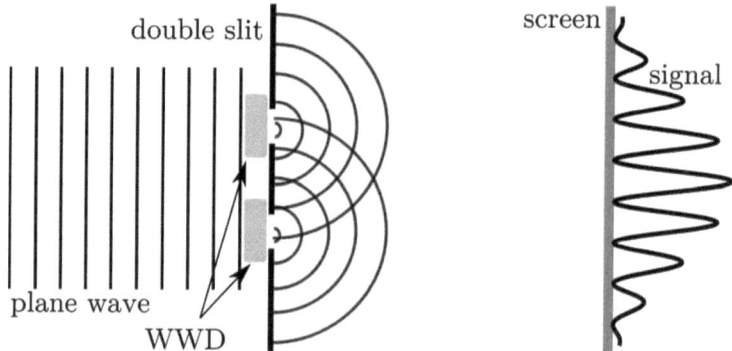

Figure 4.5: Simplified scheme of a Young-type double-slit interferometer, in which a WWD is inserted (either directly before or right after the slits) in order to measure the path of the quanton. The signal of the quanton is measured with a suitable screen in the far field.

the quanton emerging from the two slits is given by the ρ_Q from Eq. (4.4a) instead, which has to be used in place of $|\psi^{(i)}\rangle$, so that

$$P(\varphi) = \langle \psi_\varphi | \rho_Q | \psi_\varphi \rangle = \frac{1}{2}(1 + \mathcal{V}\cos(\varphi - \varphi_0)), \tag{4.18}$$

resulting in an interference pattern on the screen in which the visibility $\mathcal{V}$ and the phase offset φ_0 are found to be given by Eq. (4.16) and Eq. (4.17), respectively.

The description of a Young-type interferometer differs slightly from that of a generic two-way interferometer because, strictly speaking, it is not a two-way interferometer but instead allows for *two paths to every point on the screen*. Still, it is not surprising that the results equal those of the Mach-Zehnder interferometer; the whole formalism gives the same result as if we look at only one output port of the MZI or, vice versa, a MZI-type interferometer can be constructed from a Young-type interferometer by guiding the output of the two slits on a beam merger and monitoring its output ports on changing the path length difference from either slit to the beam merger. However, despite their relation, there still exists a fundamental difference between the two interferometers: In the MZI one can set a fixed interferometric phase φ but the output port is unpredictable and follows the random distribution from Eq. (4.15). By contrast, in the Young-type interferometer there is only one output port, but the interferometric phase that the quanton acquires is unpredictable and follows the random distribution from Eq. (4.18). We will see in Sec. 4.4 why this difference is important for us.

4.3 Phase-dependent Wave-Particle Duality [20]

4.3.1 The Two-Atom interferometer: WW Information in First-Order Interference Effects

With these tools at hand, let us start to investigate quantitatively the wave-particle duality in a system which we already know from the previous chapter: two single Λ-level atoms excited by a laser. Eichman *et al.* already observed interference fringes in such a setup by use of weak cw-excitation [89] proving thereby that it can be considered as a Young-type interferometer. In this interferometer, the atoms act as double slit and WWD at the same time, because, as we will see, their internal state can contain information on whether or not they scattered a photon.

Let us assume first that we have a far-field detection setup similar to the one discussed in Sec. 3.3 (cf. Fig. 3.3), except that we use only one single photon detector whose polarization filter is in addition fixed to transmitting σ^--polarized photons only, so that we are effectively dealing with an atomic two-level system.[4] In contrast to what was discussed in the previous chapter, we do not excite the atoms fully, but we assume that a laser pulse takes the atoms from the initial state $|+\rangle$ into the well defined superposition $\alpha|+\rangle + \beta|e\rangle$, so that their combined state is given by

$$|\chi^{(i)}\rangle = \alpha^2|++\rangle + \alpha\beta|+e\rangle + \alpha\beta|e+\rangle + \beta^2|ee\rangle. \tag{4.19}$$

The general detection operator from Eq. (3.7) describing the detection of a single photon then simplifies to

$$\hat{D}(\varphi) = |+\rangle_{a_1}\langle e| + e^{i\varphi}|+\rangle_{a_2}\langle e|, \tag{4.20}$$

while, according to Eq. (3.9), the detection probability and thus the first-order interference pattern is proportional to

$$P(\varphi) = A\langle\chi^{(i)}|\hat{D}^\dagger(\varphi)\hat{D}(\varphi)|\chi^{(i)}\rangle \propto 1 + \mathcal{V}\cos\varphi, \tag{4.21}$$

with the visibility given by

$$\mathcal{V} = |\alpha|^2. \tag{4.22}$$

This outcome can be understood a follows: Assuming that the photon is scattered by either of the atoms and thus has taken either path a or path b, we know that the atom

---

[4]We might as well employ atoms with an internal two-level structure in which case also a polarization filter in front of the detector would not be necessary. However, for the sake of not having to introduce another system, we stick to two Λ-level atoms.

which has scattered the photon must be in its ground state while the other remains in its initial superposition of states. Therefore, we can write the two possible states of the WWD from Eq. (4.3) as

$$|\chi_a\rangle = \alpha|++\rangle + \beta|+e\rangle \quad \text{and} \tag{4.23a}$$

$$|\chi_b\rangle = \alpha|++\rangle + \beta|e+\rangle. \tag{4.23b}$$

The visibility as given by Eq. (4.22) then follows immediately from Eq. (4.16).

After determining the basic properties of the interferometer, let us now turn to the question how much WW information can be extracted from the atoms. The state of the complete system after one photon has been scattered but before it is detected can be written in analogy to Eq. (4.3) as

$$|\Psi^{(f)}\rangle = \tfrac{1}{\sqrt{2}}\left(|\psi_a\rangle|\chi_a\rangle + |\psi_b\rangle|\chi_b\rangle\right), \tag{4.24}$$

where $|\psi_{a,b}\rangle$ describes the wavefunction of the photon being scattered from the respective atom.

We can now evaluate the WW knowledge one gains by reading out the two atoms. In a first attempt, we opt for a natural choice and check for both atoms whether they are in state $|e\rangle$ or $|+\rangle$, i.e., we read out the observable $\hat{N}$ with the eigenbasis $N = \{|++\rangle, |+e\rangle, |e+\rangle\}$ from the WWD; we will call this the natural observable and the natural basis, respectively. Obviously, all WW knowledge is erased if we find the atoms in state $|++\rangle$, while full WW knowledge becomes available if we find either atom in the excited state because that atom cannot have scattered the photon.[5] This also fixes the values of the q_i (cf. Eq. (4.6)) to

$$q_{++} = 0.5 \quad \text{and} \quad q_{+e} = q_{e+} = 1, \tag{4.25}$$

while the probability p_i (cf. Eq. (4.5)) to find the WWD in any eigenstate if the state of the whole system is described by Eq. (4.24) equals

$$p_{++} = |\alpha|^2 \quad \text{and} \quad p_{+e} = p_{e+} = \frac{|\beta|^2}{2}. \tag{4.26}$$

According to Eq. (4.8), $\mathcal{K}_N$ is then given by

$$\mathcal{K}_N = |\beta|^2, \tag{4.27}$$

which is less then $\sqrt{1-\mathcal{V}^2}$, the limiting value which could be reached according to Ref. [18].

---

[5] For this reason, one may consider reading out this observable a "mixture" between a quantum eraser scheme and the optimal which-way detection scheme.

4.3. PHASE-DEPENDENT WAVE-PARTICLE DUALITY

Therefore, we turn to the canonical observable: diagonalizing the operator $\hat{O}$ from Eq. (4.9) for the present WWD leads to the canonical basis with the three vectors

$$|E_0\rangle = \sqrt{\frac{1-\mathcal{V}}{1+\mathcal{V}}} \left(|++\rangle - \frac{\alpha^*}{\beta^*}(|+e\rangle + |e+\rangle)\right),$$
$$|E_a\rangle = \sqrt{\frac{\mathcal{V}}{1+\mathcal{V}}} \left(|++\rangle + \omega_+|+e\rangle + \omega_-|e+\rangle\right), \text{ and} \quad (4.28)$$
$$|E_b\rangle = \sqrt{\frac{\mathcal{V}}{1+\mathcal{V}}} \left(|++\rangle + \omega_-|+e\rangle + \omega_+|e+\rangle\right),$$

with $\omega_\pm = \frac{1-\mathcal{V}\pm\sqrt{1-\mathcal{V}^2}}{2\alpha\beta^*}$.

With these basis vectors, one finds that according to Eq. (4.5) the probability p_i of finding a certain state is given by

$$p_{E_0} = 0 \text{ and } p_{E_a} = p_{E_b} = \frac{1}{2}, \quad (4.29)$$

while according to Eq. (4.6) the probability q_i of guessing the way correctly from that state can be expressed as

$$q_{E_0} = \frac{1}{2} \text{ and } q_{E_a} = q_{E_b} = \frac{1}{2}(1+\sqrt{1-\mathcal{V}^2}). \quad (4.30)$$

Using Eq. (4.8), the WW knowledge is then calculated to be

$$\mathcal{K}_E = \sqrt{1-\mathcal{V}^2}. \quad (4.31)$$

As expected, $\mathcal{K}_E$ fulfills Eq. (4.10) optimally, i.e. $\mathcal{K}_E^2 + \mathcal{V}^2 = 1$. Thus, in general, if one aims to maximize the WW information obtained by reading out a single observable, it would be preferable to choose $\hat{E}$ over $\hat{N}$.

However, the approach as laid out so far takes into account only the correlations between the photons and the atoms to make statements about which atom scattered the photon. Not accounted for is that the entangled state in Eq. (4.3) can also lead to correlations between the detection probabilities of certain eigenstates of an observable of the atoms and the detection positions of the photon. To elucidate this point, let us investigate how much we can know about the path of photons being diffracted towards a certain position on the screen, i.e., for photons having acquired a certain phase shift φ. We know from the previous chapter that if a photon is detected with a phase shift φ the state of the atoms is projected into the state

$$|\chi_\varphi\rangle = \hat{D}(\varphi)|\chi^{(i)}\rangle = \frac{1}{\sqrt{2(1+\mathcal{V}\cos\varphi)}} \left[\alpha(1+e^{i\varphi})|++\rangle + \beta|+e\rangle + e^{i\varphi}\beta|e+\rangle\right]. \quad (4.32)$$

If we now look at the WW knowledge we can extract from the atoms by reading out the natural observable, we see that the probability p_i of finding certain basis states depends on φ, where the p_i are given by

$$p_{++} = \mathcal{V}\frac{1+\cos\varphi}{1+\mathcal{V}\cos\varphi} \quad \text{and} \quad p_{e+} = p_{+e} = \frac{1-\mathcal{V}}{2(1+\mathcal{V}\cos\varphi)}. \tag{4.33}$$

This in turn leads to the phase-dependent WW knowledge

$$\mathcal{K}_N(\varphi) = \frac{1-\mathcal{V}}{1+\mathcal{V}\cos\varphi}.^6 \tag{4.34}$$

It is interesting to note that the WW knowledge is now indeed correlated with the position on the screen where the quanton is detected. However, before we start to investigate this result in more detail, let us find out whether it is a peculiarity of the specific setup considered here or whether we can also find it in other interferometers.

4.3.2 The Two-Atom interferometer: WW Information in Second-Order Interference Effects

In the previous section, the most simple type of interferometer, namely an interferometer where interference pattern was displayed in the intensity distribution of the light scattered by two atoms, was considered. Let us now move to an interferometer which is even more closely related to the one discussed in Sec. 3.3. Again, two Λ-level atoms constitute the interferometer. However, the photons scattered by the two atoms are now observed with two polarization sensitive detectors D_1 and D_2, where the polarization filter ε_1 in front of D_1 is fixed to transmit σ^--polarized photons only, while the second polarization filter ε_2 is oriented in an arbitrary direction described by the vector $(\varepsilon_-^*, \varepsilon_+^*)$. As in Sec. 3.3, the signal is given by the events in which both detectors register a photon. As we have seen in Eq. (3.23), there occurs an interference pattern in the second-order intensity correlation function depending on the position of the two detectors and the relative orientation of the two polarization filters.

The detection operators describing the detection of the photons, being a special case of Eq. (3.10), are given by

$$\hat{D}_1(\varphi_1) = \varepsilon_-|+\rangle_{a_1}\langle e| + e^{i\varphi_1}\varepsilon_-|+\rangle_{a_2}\langle e| \quad \text{and} \tag{4.35a}$$

$$\hat{D}_2(\varphi_2) = \varepsilon_-|+\rangle_{a_1}\langle e| + \varepsilon_+|-\rangle_{a_1}\langle e| + e^{i\varphi_2}(\varepsilon_-|+\rangle_{a_2}\langle e| + \varepsilon_+|-\rangle_{a_2}\langle e|). \tag{4.35b}$$

The interference pattern can therefore be written in analogy to Eq. (3.23) as

$$P(\varphi) \propto \langle ee|\hat{D}_1^\dagger(\varphi_1)\hat{D}_2^\dagger(\varphi_2)\hat{D}_2(\varphi_2)\hat{D}_1(\varphi_1)|ee\rangle \propto 1 + \mathcal{V}\cos\varphi_{21} \tag{4.36}$$

---

[6] If we adopt the same procedure for the canonical observable $\hat{E}$, we find that it is not correlated with φ, i.e., $\mathcal{K}_E(\varphi) = \mathcal{K}_E$.

4.3. PHASE-DEPENDENT WAVE-PARTICLE DUALITY

with $\mathcal{V} = |\boldsymbol{\varepsilon}_1 \cdot \boldsymbol{\varepsilon}_2|^2 = |\varepsilon_-|^2$ and $\varphi_{21} = \varphi_2 - \varphi_1$.

This interference pattern results from the superposition of the two quantum paths corresponding to the two possibilities that either D_1 detects the photon scattered by a_1 and D_2 registers the photon emitted by a_2 or vice versa. The two WWD states describing these two possibilities are given by

$$|\chi_a\rangle = \varepsilon_-|++\rangle + \varepsilon_+|+-\rangle \quad \text{and} \quad (4.37a)$$

$$|\chi_b\rangle = \varepsilon_-|++\rangle + \varepsilon_+|-+\rangle, \quad (4.37b)$$

while the final state of the two atoms after detection of both photons is given by

$$|\chi_{\varphi_{21}}\rangle = \hat{D}_2(\varphi_2)\hat{D}_1(\varphi_1)|ee\rangle = \frac{1}{\sqrt{2(1+\mathcal{V}\cos\varphi_{21})}}\left[\varepsilon_-(1+e^{i\varphi_{21}})|++\rangle + \varepsilon_+|+-\rangle + e^{i\varphi_{21}}\varepsilon_+|-+\rangle\right]. \quad (4.38)$$

As we can see, this case has a one-to-one correspondence with the case of first-order interference discussed in the previous section if we replace in the state equations φ with φ_{21}, $|e\rangle_{a_{1,2}}$ with $|-\rangle_{a_{1,2}}$, and (α, β) with $(\varepsilon_-, \varepsilon_+)$. Therefore, all results regarding wave-particle duality, and in particular Eq. (4.34) for a phase-dependent WW knowlegde, also hold for the case of an interferometer displaying an interference pattern in the second-order intensity correlation.

4.3.3 General Interferometer

So far, we have seen two examples in which the effects discussed in Sec. 4.3.1 are displayed. Let us now turn to a general two-way interferometer with a WWD. We will use the Mach-Zehnder formalism, as it allows for a deliberate choice of the phase shift and includes all results for a Young-type interferometer (while the reverse is not true). The states of the WWD, after the quanton has passed either the upper path or the lower path, are represented in a very general manner by $|\chi_a\rangle$ and $|\chi_b\rangle$ and the visibility of the interference pattern is given by the overlap of those two final states (cf. Eq. (4.16)). We have detailed in Sec. 4.2.2 how to evaluate the WW information in the general case. However, we expect now to find also a correlation between φ and $\mathcal{K}$ which has not been treated above. In order to do so, we start by noting that according to Eq. (4.14) the final state of the WWD after detection of the quanton can be written as

$$|\chi_\varphi^\pm\rangle = |\chi_a\rangle \pm e^{i(\varphi-\varphi_0)}|\chi_b\rangle, \quad (4.39)$$

88   CHAPTER 4. COMPLEMENTARY MEASUREMENTS

where the sign depends on the output port in which the quanton is found.[7] Obviously, the identity

$$|\chi_\varphi^-\rangle = |\chi_{\varphi+\pi}^+\rangle \tag{4.40}$$

always holds. In order to evaluate the WW knowledge in dependence on the phase shift, we first need to choose an observable $\hat{W}$ which is to be read out from the WWD. While for each of its eigenstates $|W_k\rangle$ the probability q_i of guessing the way correctly does not depend on φ (cf. Eq. (4.6)), the probability $p_i^\pm(\varphi)$ of finding the quanton with a certain phase shift in a particular output port and the WWD in the final state $|W_i\rangle$ certainly does. In analogy to Eq. (4.5), we can write $p_i^\pm(\varphi)$ as

$$p_i^\pm(\varphi) = P_\varphi(|\psi^\pm\rangle) \cdot |\langle W_i|\chi_\varphi\rangle|^2, \tag{4.41}$$

so that in equivalence to Eq. (4.8) we write $\mathcal{K}_W^\pm(\varphi)$ as

$$\mathcal{K}_W^\pm(\varphi) = 2\frac{\sum_i p_i^\pm(\varphi) q_i}{\sum_i p_i^\pm(\varphi)} - 1, \tag{4.42}$$

where the denominator takes care of the fact that the conditional probabilities do not sum up to one as long as we do not know for sure through which output port the quanton exits the interferometer.

If we want to refind the effects discovered in the interferometers above, we need an observable whose eigenstates either carry full WW information or none at all. As proved in App. B, one can always decompose the Hilbert space spanned by $|\chi_a\rangle$ and $|\chi_b\rangle$ into three vectors $|a\rangle$, $|b\rangle$, and $|0\rangle$, such that $|a\rangle$ and $|b\rangle$ carry full WW information while $|0\rangle$ carries none. With such a decomposition, Eq. (4.39) can be rewritten in a form equivalent to Eqs. (4.32) and (4.38)

$$|\chi_\varphi^\pm\rangle = c_0(1 \pm e^{i(\varphi-\varphi_0)})|0\rangle + c_1|a\rangle \pm c_1 e^{i\varphi}|b\rangle, \tag{4.43}$$

where $c_0 = \sqrt{\mathcal{V}}$ and $c_1 = \sqrt{1-\mathcal{V}}$ (cf. App. B). Since this is formally equivalent to the previous sections, we will call the observable defined by the eigenbasis ($|0\rangle$, $|a\rangle$, $|b\rangle$) the natural observable, even though it may not be, in general, the most intuitiv or simple choice. In addition, because of this formal equivalence, all results concerning the phase-dependent WW information carry over to the most general case. In particular, for the canonical basis, the phase-dependent WW knowlegde can be written in analogy

---

[7]Here, and in all following equations, a Young-type interferometer as discussed in the examples above is described by considering the positive case only. Therefore, if not specified otherwise, we will always discuss the general interferometer as a Mach-Zehnder interferometer, keeping in mind that the results with the positive superscript all hold also for a Young-type interferometer.

4.3. PHASE-DEPENDENT WAVE-PARTICLE DUALITY

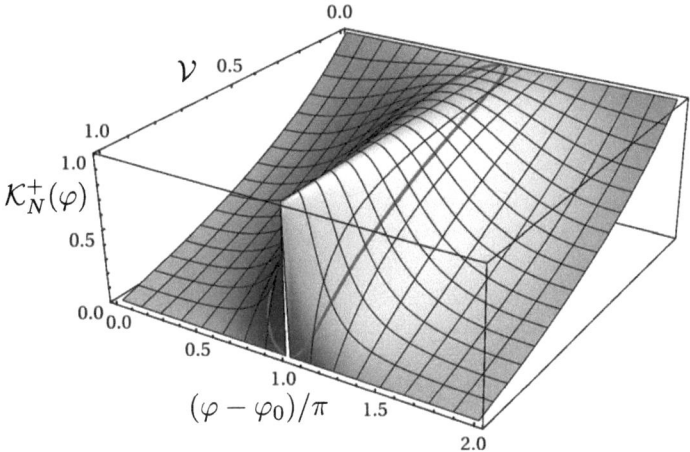

Figure 4.6: $\mathcal{K}_N(\varphi)$ plotted as a function of the phase shift φ (in units of π) and the visibility $\mathcal{V}$. The thick red line marks the region in which the natural observable allows to gain more WW information from the WWD than the canonical observable.

to Eq. (4.31) as

$$\mathcal{K}_E^\pm(\varphi) = \mathcal{K}_E^\pm = \mathcal{K}_E = \sqrt{1-\mathcal{V}^2} \qquad (4.44)$$

while for the natural basis, the phase-dependent WW knowledge is given in equivalence to Eq. (4.34) as

$$\mathcal{K}_N^\pm(\varphi) = \frac{1-\mathcal{V}}{1 \pm \mathcal{V}\cos(\varphi-\varphi_0)}. \qquad (4.45)$$

4.3.4 Physical Interpretation of Phase-dependent WW information

In order to show how the WW knowledge correlates with the interferometric phase, $\mathcal{K}_N^+$ as given in Eq. (4.45) is plotted in Fig. 4.6. Noteworthy is in particular that for all quantons arriving in the minima of the interference pattern $P_\varphi(|\psi^+\rangle)$, i.e., at $\cos\varphi - \varphi_0 = -1$, one always has full WW knowledge, regardless of the value of the visibility (cf. Fig. 4.6).[8] Moreover, while the observable $\hat{E}$ performs better on average, for all quantons arriving in a region around those minima, the observable $\hat{N}$ provides more WW information than $\hat{E}$. By comparing Eq. (4.44) and Eq. (4.45), one finds that

---

[8]Except if $\mathcal{V} = 1$, in which case $\mathcal{K}_N^+ = \frac{0}{0}$ is not defined. This is however not a problem: The case $\mathcal{V} = 1$ and $\cos(\varphi - \varphi_0) = -1$ is unphysical, since there is a vanishing probability to detect a quanton in the minimum of an interference pattern with perfect visibility, i.e., the WWD will never be read out for this set of values (cf. Eq. (4.2)).

this is true for quantons for which the phase shift φ fulfills the following relation:

$$\cos(\varphi - \varphi_0) < \frac{\sqrt{\frac{1-\mathcal{V}}{1+\mathcal{V}}} - 1}{\mathcal{V}}. \qquad (4.46)$$

One way of understanding these correlations is by looking at the WWD wavefunction upon detection of a quanton at the output of the interferometer. After the detection of a quanton with phase shift φ, one finds the state of the WWD to be projected into the state given by Eq. (4.43). From this equation, we can see that whenever a quanton is detected in state $|\psi^+\rangle$ in the vicinity of $\cos(\varphi - \varphi_0) = -1$, there is only a tiny chance of finding the WWD in the state $|0\rangle$, whereas one is almost certain to find the WWD either in state $|a\rangle$ or $|b\rangle$. This is a result of an interference effect in the WWD which cancels the overlapping parts of $|\chi_a\rangle$ and $|\chi_b\rangle$, so that only the non-overlapping parts, $c_1|a\rangle$ and $c_1|b\rangle$, contribute to the final WWD wavefunction, both determining unambigouosly the path of the quanton.

We note that even though the argumentation in the previous section are based on the state $|\chi_\varphi^\pm\rangle$, created in the WWD upon detection of a quanton, all results are in fact *independent* of the order in which the quanton and the WWD are read out, since only the joint probability of detecting a quanton for a specific value of φ in a certain state and finding the WWD in a particular state enters into the equations. In order to elucidate this point, it may help to turn around the argumentation for the case of the natural observable. If we rewrite the state of the quanton-WWD system from Eq. (4.13) by use of the natural basis in the form

$$|\Psi^{(f)}\rangle = c_0(|\psi_a\rangle + e^{i(\varphi-\varphi_0)}|\psi_b\rangle)|0\rangle + c_1|\psi_a\rangle|a\rangle + c_1 e^{i(\varphi-\varphi_0)}|\psi_b\rangle|b\rangle, \qquad (4.47)$$

one can clearly see from this equation that the detection of the WWD in either state $|a\rangle$ or $|b\rangle$ projects the quanton into state $|\psi_a\rangle$ or $|\psi_b\rangle$, respectively. Thus, the quanton is emerging exactly from one slit, in which case no interference occurs and one has an equal probability of detecting the quanton in any output port for all φ. However, if the WWD is found to be in state $|0\rangle$, the wavefunction of the quanton is projected into the state $\frac{1}{\sqrt{2}}(|\psi_a\rangle + e^{i(\varphi-\varphi_0)}|\psi_b\rangle)$, i.e., an equal superposition of both paths, in which case all WW information has been erased. Thus, with the WWD being found in $|0\rangle$, no quantons arrive in the minima of the interference pattern $P_\varphi(|\psi^\pm\rangle)$ and only very few in their vicinity around $\cos(\varphi - \varphi_0) = \mp 1$ due to destructive interference between $|\psi_a\rangle$ and $|\psi_b\rangle$. Therefore, all of the quantons that actually are detected in the minima of an interference pattern with $\mathcal{V} < 1$ and most of the ones in the surroundings coincide with a WWD state which carries full WW information.

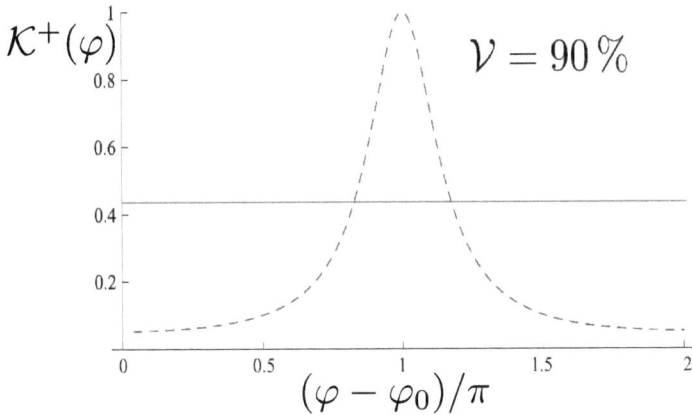

Figure 4.7: WW knowledge which is extracted from the WWD when reading out either the natural observable (dashed purple line) or the canonical observable (full olive line) for a visibility of $\mathcal{V} = 0.9$.

4.4 Improving the WW Knowledge by a Delayed Choice of the WWD Observable

We have seen in Sec. 4.3.4 that the natural observable performs better than the canonical observable for quantons arriving close to the minima of the interference pattern, while already in Sec. 4.3.1 it was shown that on average over a large number of experimental runs it performs worse. This property can be put to use by the following simple strategy. One starts by measuring the quanton and finds it under a certain phase shift and at specific output port. Since for both output ports and all phase shifts φ, we know how much WW information is to be expected for either of the two observables, we can *after having measured the quanton* choose the observable which performs better for this specific output port and value of φ. Let us exemplify this strategy for a visibility of $\mathcal{V} = 0.9$. In Fig. 4.7, $\mathcal{K}_E^+$ and $\mathcal{K}_N^+$ are plotted for this visibility. Whenever a quanton arrives in the region around φ where the dashed purple line is higher than the full olive line (cf. also Eq. (4.46)), one chooses to read out the natural observable from the detector, while in all other cases one opts for the canonical observable. Following this strategy will be denoted as obtaining the WW knowledge $\mathcal{K}_{N,E}^\pm$.

This strategy is obviously advantageous compared to reading out the same observable for every run of the experiment. However, these results cannot be interpreted directly in terms of quantitative wave-particle duality because each result is obtained only in a single shot of the experiment. Consider for example the case in which we have $\cos(\varphi - \varphi_0) = -1$. If we find the quanton in the state $|\psi^+\rangle$, we obviously have full WW information (cf. Fig. 4.7). However, most of the time, the quanton will be found in the state $|\psi^-\rangle$, in which case less WW information is accessible. Therefore, we cannot claim that full WW information is available. Rather, in an arbitrary run, the output port is unpredictable and follows the distribution from Eq. (4.15). Therefore, let us take a look on the average WW knowlegde that is obtained in an arbitrary run of a Mach-Zehnder interferometer for a fixed phase φ. It can be seen from Eq. (4.40) that the curve for the second output port is necessarily the same as the one for the first, only shifted by π. Therefore, by averaging over the two possible output ports, $\langle \mathcal{K}_{N,E}\rangle(\varphi)$ is given as the average of $\mathcal{K}_{N,E}(\varphi)$ and $\mathcal{K}_{N,E}(\varphi + \pi)$, weighted with the respective probabilities of finding the quanton at the first or second output port

$$\begin{aligned}\langle \mathcal{K}_{N,E}\rangle(\varphi) &= P_\varphi(|\psi^+\rangle) K^+_{N,E}(\varphi) + P_\varphi(|\psi^-\rangle) K^-_{N,E}(\varphi) \\ &= P_\varphi(|\psi^+\rangle) K^+_{N,E}(\varphi) + P_{\varphi+\pi}(|\psi^+\rangle) K^+_{N,E}(\varphi + \pi).\end{aligned} \quad (4.48)$$

Using $\langle \mathcal{K}_{N,E}\rangle(\varphi)$, we obtain a local description of the available WW knowledge, since the value of $\langle \mathcal{K}_{N,E}\rangle(\varphi)$ depends on φ. Therefore, also the relation to $\mathcal{V}$ and consequently quantitative wave-particle duality can be described in a localized, i.e. φ-dependent, form.

As we want to compare the WW knowledge to the visibility which is a global quantity in the sense that it cannot be defined for a single value of φ, but is only obtained if the whole parameter range for φ is scanned, we can also introduce the global description of WW knowledge $\langle \mathcal{K}_{N,E}(\varphi)\rangle$ as an average over all values of φ. We then arrive at the phase-independent equation

$$\langle \mathcal{K}_{N,E}(\varphi)\rangle = \frac{1}{2\pi} \int_0^{2\pi} \langle \mathcal{K}_{N,E}\rangle(\varphi)\, d\varphi, \quad (4.49)$$

which gives a global description of the WW knowledge and will therefore also enable a global description of wave-particle duality. Both, the well-known global description as well as the new local description of duality are well justified. Therefore, both will be investigated in the following.

We note that one could try to argue that an average over the output ports is only necessary in a Mach-Zehnder interferometer, while, in a Young-type interferometer there exists only one output port and consequently we do have full WW knowledge for a certain phase. However, this argument ignores the fact that the phase itself is randomly distributed according to Eq. (4.18) and therefore, the average over the phase must be taken right away, since this is the randomly distributed quantity.[9] Since we know that the Young-type interferometer is described by considering only one output

---
[9] For this reason, there exists only a global description of wave-particle duality in Young-type interferometers.

port of a Mach-Zehnder interferometer and obviously an average over a full 2π interval of only one summand of Eq. (4.48) will give the same value as the average taken over the whole sum, it can be concluded that Eq. (4.49) also describes the average WW knowledge in a Young-type interferometer.

It is also interesting to note that the values obtained for $\langle \mathcal{K}_{N,E}(\varphi) \rangle$ will, for all $0 < \mathcal{V} < 1$, violate Eq. (4.10), since all values of $\mathcal{K}_{N,E}^{\pm}$ are at or above $\mathcal{K}_E = \langle \mathcal{K}_E \rangle(\varphi) = \langle \mathcal{K}_E(\varphi) \rangle$, the limit set by this equation. Analytic results for $\langle \mathcal{K}_{N,E} \rangle(\varphi)$ and $\langle \mathcal{K}_{N,E}(\varphi) \rangle$ are easily calculated according to the given formulas. However, as explained in the next section, there exists a slightly better strategy than the one presented here. Therefore, we will not give explicit results here, as they are very similar but slightly less optimal than the ones discussed in the next section.

4.5 Further Improving the WW Knowledge [21]

By choosing the better of the natural and the canonical observable, we have been able to improve on the WW knowledge extracted from the WWD. However, there exists of course an infinite number of observables, given by an arbitrary basis spanning the Hilbert space of the WWD, and it is by no means obvious why no other observable should be able to perform even better for certain values of φ. In order to find the optimal observable, Eq. (4.42) has to be optimized with regard to the basis W and in dependence on φ. Since in this case, the basis is fixed by the optimization process, we will denote the WW knowledge obtained in this way simply by $\mathcal{K}$ (without subscript). Unfortunately, no analytic solution to that optimization problem has been found. Therefore, Monte Carlo simulations were performed to obtain a numeric approximation of the solution. For these simulations, the parameter space of Eq. (4.42) was sampled at 50 evenly spaced values of φ and for visibilities $\mathcal{V} \in \{0.01, 0.22, \ldots, 0.98, 0.99, 0.999\}$. For each combination of these parameters, $\mathcal{K}_W$ was calculated for 10,000 randomly chosen orthonormal sets of basis vectors $\{W\}$ and the highest value was retained as the optimal one for these parameters. The results were stable in the sense that neither for a higher number of random bases nor for different runs of the simulations, the results varied noticably.

The result of one run for a visibility of $\mathcal{V} = 0.9$ is shown in Fig. 4.8. One can see from the graphical representation that the two observables which have been discussed above, namely the canonical and the natural observable $\hat{E}$ and $\hat{N}$, respectively, actually do give a good approximation of the best possible result. Only close to the region where both observables give a comparable amount of WW information, there exist other observables which perform noticably better.

With these optimized values we can again take a look at the maximally available averaged WW knowledge $\langle \mathcal{K} \rangle(\varphi)$ (cf. Eq. (4.48)) and $\langle \mathcal{K}(\varphi) \rangle$ (cf. Eq. (4.49)). The

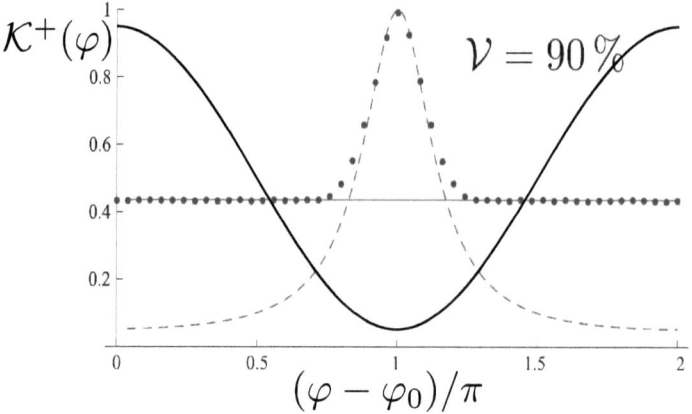

Figure 4.8: The results of a numerical calculation of the optimal value of $\mathcal{K}_W$ in dependence of φ for $\mathcal{V} = 0.9$. The 51st data point at $\varphi = 2\pi$ is only a copy of the data point at $\varphi = 0$, since all equations are 2π-periodic. The black line is a plot of the interference pattern as given for $P_\varphi(|\psi^+\rangle)$ in Eq. (4.15).

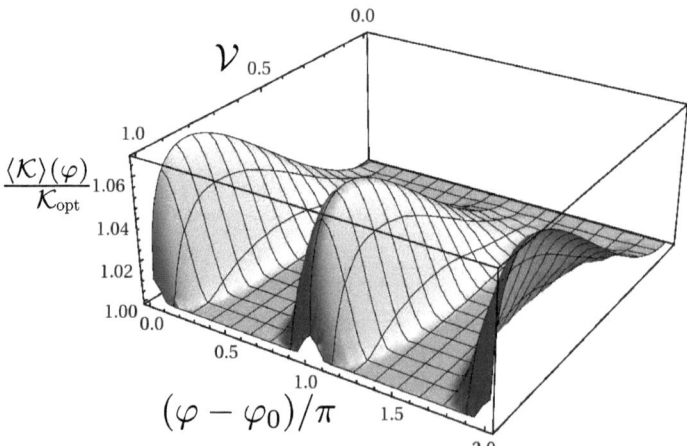

Figure 4.9: Plot of the numerical results for the WW knowledge $\langle \mathcal{K} \rangle(\varphi)$ in dependence on the visiblitiy $\mathcal{V}$ and the interferometric phase φ. Plotted is the relative excess of WW knowledge compared to the limit $\mathcal{K}_{\text{opt}} = \sqrt{1 - \mathcal{V}^2}$ set by Eq. (4.10).

4.6. DO WE STILL OBTAIN WW INFORMATION?

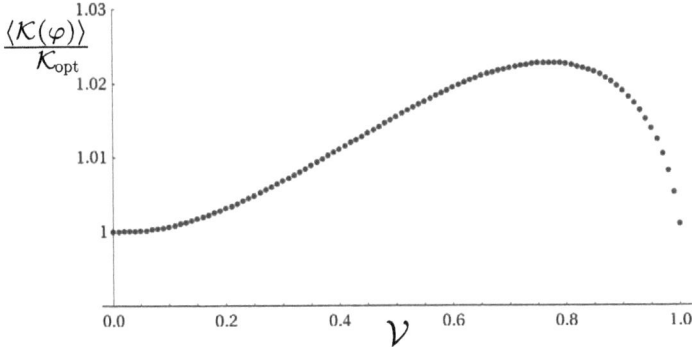

Figure 4.10: Plot of the numerical results for the average WW knowledge $\langle \mathcal{K}(\varphi) \rangle$. As in Fig. 4.9, $\mathcal{K}_{\text{opt}} = \sqrt{1 - \mathcal{V}^2}$ (cf. Eq. (4.10)).

results are shown in Fig. 4.9 and Fig. 4.10, respectively.

Considering a local description of wave-particle duality, the relative violation of Eq. (4.10) by $\langle \mathcal{K} \rangle(\varphi)$ depends on φ and $\mathcal{V}$ and is the strongest for $\mathcal{V} \approx 0.84$ for which the WW knowledge exceeds the value allowed by Eq. (4.10) by almost seven percent for φ an integer multiple of π. Therefore, we can introduce a local duality relation which is approximated by

$$\langle \mathcal{K} \rangle(\varphi)^2 + \mathcal{V}^2 \lesssim 1,040 \tag{4.50}$$

Furthermore, we find that for $\langle \mathcal{K}(\varphi) \rangle$ the strongest violation of Eq. (4.10) occurs for $\mathcal{V} \approx 0.77$ and reaches more than 2 % so that we may approximate a new global duality relation by

$$\langle \mathcal{K}(\varphi) \rangle^2 + \mathcal{V}^2 \lesssim 1,025. \tag{4.51}$$

Note that the relative violation remains quite small for both the local and the global form, because the higher the visibility, the less likely it becomes to detect a quanton for which the amount of expected WW knowledge exceeds the limit set by Eq. (4.10) noticably (cf. Fig. 4.8).

4.6 Do We Still Obtain WW Information?

One can raise the valid question whether in our scenario we do actually measure WW information since after the readout of the quanton and the corresponding projection, the WW information cannot be verified by subsequently measuring the path of the quanton (according to Fig. 4.3). We answer this question to the positive, since during its interaction with the WWD the quanton deposits WW information in the WWD and this WW information persists no matter what happens with the quanton thereafter [155]. Thus, a projection of the quanton (and obviously also the experimenter's choice of the basis with which he plans to read out the WWD) does not change the information content of the WWD with respect to the path of the quanton; the projection only reflects our increased knowledge about the state of the WWD. Therefore, reading out the state of the WWD will indeed deliver information about the path of the quanton, independent of whether the quanton has been read out or not, i.e., independent of whether we describe the state of the WWD by a partial trace over Eq. (4.3) or by Eq. (4.43) [155, 156]. Hereby, the amount of WW information that is stored in each basis state $|W_k\rangle$ of the observable $\hat{W}$ chosen to read the WWD, i.e., the degree of correlation between the outcome of a measurement of the WWD and the way actually taken inside the interferometer, can be experimentally tested in the setup shown in Fig. 4.3. This holds true regardless of whether the quanton or the WWD is read out first.

4.7 Example: The Micromaser

The outcomes described in the previous sections can be found in all interferometers. So far, we have investigated photonic interferometers using two atoms as photon scatterers. These have several disadvantages, the most important being that the beam splitter is not seperated from the WWD.[10] Therefore, as a specific quantum optical example which has less drawbacks, let us look at a variation of the famous quantum eraser setup by Scully et al. [137], shown in Fig. 4.11. Disregarding the laser-pulse, the rf-pulse and the cavities from Fig. 4.11 for the moment, the setup forms a basic atom interferometer: A plane atomic wave hits a set of wider slits where the beam is collimated and then illuminates two narrow slits where the wave is diffracted. The wavefunction describing the center of mass motion of the atom after the second double slit is then given by a sum of the two wavefunctions originating at the two narrow slits:

$$\Psi(\mathbf{r}) = \frac{1}{\sqrt{2}}[\psi_1(\mathbf{r}) + \psi_2(\mathbf{r})]. \qquad (4.52)$$

The probability to detect an atom at a certain position $\mathbf{r}$ on the screen is then given by:

$$P(\mathbf{r}) = \frac{1}{2}\left[|\psi_1(\mathbf{r})|^2 + |\psi_2(\mathbf{r})|^2 + \psi_1^*(\mathbf{r})\psi_2(\mathbf{r}) + \psi_2^*(\mathbf{r})\psi_1(\mathbf{r})\right]. \qquad (4.53)$$

---

[10]For example, this fact makes it impossible to realize an arbitrary asymmetric WWD, e.g., a WWD placed only in one of the two arms of the interferometer.

The interference is given as usual by the cross terms $\psi_1^*\psi_2 + \psi_2^*\psi_1$. Now we include a laser which excites a long lived Rydberg state $|x\rangle$ in the atoms and an rf-pulse which transfers the internal state of the atom into the coherent superposition $\cos\theta|x\rangle + \sin\theta|y\rangle$ of $|x\rangle$ and an energetically higher lying Rydberg state $|y\rangle$. Behind the laser and the rf-pulse generator, we place two identical cavities C_1 and C_2, arranged as shown in Fig. 4.11. The cavities – both initially in the vacuum state $|0\rangle$ – are assumed to be tuned into resonance with the transition of the internal atomic state $|x\rangle \to |y\rangle$ and their parameters are trimmed such that exactly half a Rabi oscillation takes place for the atomic state $|y\rangle$ when an atom is passing through the cavities. As a result, after the passage of the atom through C_1 and C_2, the internal atomic state is $|x\rangle$, while the cavities are left in a superposition of either containing a photon or not, thus marking the path of the atom:

$$\Psi(\mathbf{r}) = \frac{1}{\sqrt{2}}\left[\psi_1(\mathbf{r})(\cos\theta|00\rangle + \sin\theta|10\rangle) + \psi_2(\mathbf{r})(\cos\theta|00\rangle + \sin\theta|01\rangle)\right]|y\rangle, \quad (4.54)$$

where the kets $|C_1 C_2\rangle$ denote the number of photons in the cavities C_1 and C_2, respectively. The probability to find an atom at point $\mathbf{r}$ is now given by

$$P(\mathbf{r}) = \frac{1}{2}\left[|\psi_1(\mathbf{r})|^2 + |\psi_2(\mathbf{r})|^2 + \cos^2\theta \psi_1^*(\mathbf{r})\psi_2(\mathbf{r}) + \cos^2\theta \psi_2^*(\mathbf{r})\psi_1(\mathbf{r})\right], \quad (4.55)$$

where the visibility $\mathcal{V}$ of the interference pattern is reduced to $\mathcal{V} = \cos^2\theta$. At the same time, the detection of an atom at position $\mathbf{r}$ projects the cavities into the state

$$|C_1 C_2\rangle = \cos\theta(1 + e^{i\varphi})|00\rangle + \sin\theta|10\rangle + \sin\theta e^{i\varphi}|01\rangle. \quad (4.56)$$

In this case, the natural basis is simply made up of the Fock number states $|00\rangle$, $|01\rangle$, and $|10\rangle$ of the cavities, i.e., one simply checks whether one finds a photon in either cavity. If one does find a photon, one knows that the atom has passed through the respective cavity, while if one finds both cavities to be empty, all WW information contained in the state of the WWD described by Eq. (4.56) has been erased. As discussed in Sec. 4.3.4, the natural basis provides more WW information than the canonical basis for all atoms arriving close to the minima of the interference pattern, while the canonical basis performs better over a wider range around the maxima and also on average over a large number of runs of the experiment.

4.8 Conclusion

In conclusion, we have shown that while it is not possible to choose a single observable which beats the canonical observable in terms of average WW information, the natural observable does lead to more WW information for a certain subset of quantons passing

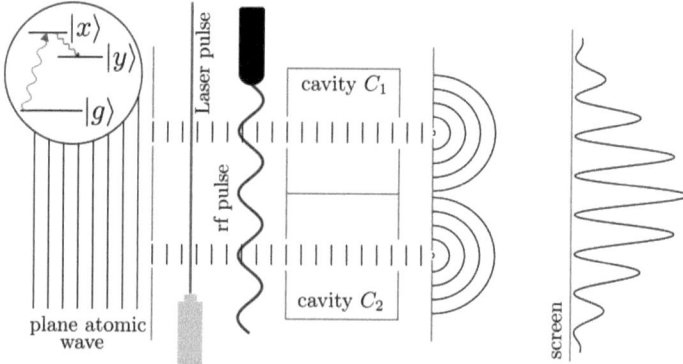

Figure 4.11: Simplified scheme of the micromaser setup: atoms enter as a plane wave from the left. Two wide appertures seperate the atomic beam spatially into two collimated beams. Then, a laser excites the atoms into a long-lived Rydberg state $|x\rangle$ and an rf-pulse transfers them from there into a superposition of two Rydberg states $\cos\theta|x\rangle + \sin\theta|y\rangle$. The cavities are tuned to the resonance $|x\rangle \to |y\rangle$ and their parameters are adjusted such that $|y\rangle$ deexcites with certainty under emission of a photon. After passing the cavities, the atoms are diffracted at two narrow slits and their interference pattern is observed on a screen.

through a Young-type interferometer. In addition and especially for high values of the visibility, there is an automatic sorting taking place: the less which-way information the WWD reveals, the more likely it is to find the quanton close to the maxima of the interference pattern; thus, we have realized what might be called a 'self-sorting quantum eraser'. Furthermore, we found that by adopting a strategy of choosing the observable which to read out from the WWD only after the quanton itself has been measured, it becomes possible to extract more WW information from the WWD than previously thought possible. Further work might include the extension of the analysis to asymmetric interferometers and mixed WWD states.

Chapter 5

Summary

In this thesis, aspects of the quantum mechanical measurement process have been investigated in three concrete applications to quantum optics.

Chapter 2 treats the measurement of the expectation values of specific operators at the example of correlation measurements of the electromagnetic field at optical frequencies. Starting from a well-known procedure to measure the variances of the quantum Stokes parameters, we introduced a new method of how to measure Nth-order correlations of a single light beam in which equal powers of the photon creation and annihilation operator appear in employing a simple setup. This procedure might be interesting in general to characterize the quantum states of the electromagnetic field and in particular it can serve as a full state tomography for photon number states.

In Chapter 3, this measurement process is generalized to arbitrary spatial modes $\mathbf{k}$, but specialized to specific sources, namely single photon sources with an internal Λ-level scheme. Thereby, the measurement process is not considered as an end in itself, i.e., to measure the signal strength of higher-order correlation functions of the emitters. Instead, the projective nature of the measurement process is used in order to create large families of quantum states among the emitters. In particular, in Sec. 3.3, we were able to show that two of those emitters can be entangled to an arbitrary degree by detecting the photons they emit with two polarization sensitive detectors in the far field of the atoms. Thereby, it turned out that for specific detector positions, the entanglement as quantified by the concurrence can be made to follow a surprisingly simple analog to Malus' Law. In Sec. 3.4 we introduced a method to generate arbitrary total angular momentum eigenstates in the two ground states of the single photon emitters by a setup which mimicks the coupling of angular momenta, while Sec. 3.5 uses a proposal by Xia et al. to allow for the generation of a class of cluster states, a family of states which is highly interesting in the field of quantum computation. Sec. 3.6 then generalizes the approach developed in the former sections to investigate whether it might be possible to generate an arbitrary quantum state in N remote qubits with the technique of projective measurements. We came to an affirmative answer and an explicit solution for the case of two or three single photon emitters, but it also became clear that there is no general solution for more than six emitters. Thus, while our approach is capable to generate wide varieties of states in an arbitrary number of qubits, the approach is not a solution to the problem of creating the "anystate". Sec. 3.7 concludes the chapter by showing that with a slightly modified setup, the same techniques can serve to generate the same families of states among photonic qubits instead of massive ones.

Finally, Chapter 4 showed how the two-atom system introduced in Chapter 3 can be interpreted as an interferometer with a which-way detector. This aspect of viewing the setup was used as a starting point to discuss in a very general manner the possibilities and limits of measurements as they apply to the well-known and much-debated aspect of wave-particle duality. We found that if a certain measurement is performed on the which-way detector in order to gain insight into which path the interfering object took inside the interferometer, then the amount of WW information one is able to extract from the WW detector is correlated with the relative phase that this object accumulates on the two paths. The most notable result is that for certain phase shifts, the WW information that can be extracted from the WW detector using a specific observable can surpass the amount of WW information one obtains when reading out the observable so far known to be the optimal one. This property is then employed in a next step to optimize the observables which to read out from the WWD in dependence on the phase shift. We show that with this strategy one can extract more which-way information than has been thought possible so far.

Appendix A

Solving the Anystate Equations for $N = 3$

In Sec. 3.6.4 it remained to be shown that the following system of non-linear equations can be solved:

$$\begin{pmatrix} 1 & 0 & 0 & 0 & 0 & 0 \\ 0 & 1 & 0 & 0 & 0 & 0 \\ 0 & 0 & 1 & 0 & 0 & 0 \\ 0 & 0 & 0 & 1 & 0 & 0 \\ 0 & 0 & 0 & 0 & 1 & 0 \\ 0 & 0 & 0 & 0 & 0 & 1 \\ 0 & 0 & 0 & 0 & 0 & 0 \\ 0 & 0 & 0 & 0 & 0 & 0 \end{pmatrix} \begin{pmatrix} \alpha_3 \\ \gamma_3 \\ \epsilon_3 \\ \beta_3 \\ \delta_3 \\ \zeta_3 \end{pmatrix} = \begin{pmatrix} x_1 \\ x_2 \\ x_3 \\ x_4 \\ x_5 \\ x_6 \\ g - \{\delta,\beta\}x_3 - \{\delta,\epsilon\}x_4 - \{\beta,\epsilon\}x_5 \\ h - \{\zeta,\delta\}x_4 - \{\zeta,\beta\}x_5 - \{\delta,\beta\}x_6 \end{pmatrix} \quad (A.1)$$

with $x_1, x_2, x_3, x_4, x_5, x_6$ being terms that arise by applying the Gaussian elimination scheme. Using in addition the notation $[x, y] = x_1 y_2 - x_2 y_1$, they can be written as

$$x_1 = \frac{-\{\alpha,\delta\}\{\alpha,\zeta\}a + \{\alpha,\delta\}\{\alpha,\epsilon\}b + \{\alpha,\gamma\}\{\alpha,\zeta\}c - \{\alpha,\gamma\}\{\alpha,\epsilon\}e}{2\alpha_1\alpha_2[\delta,\gamma][\zeta,\epsilon]}, \quad (A.2)$$

$$x_2 = \frac{-\{\beta,\gamma\}\{\gamma,\zeta\}a + \{\gamma,\epsilon\}\{\beta,\gamma\}b + \{\alpha,\gamma\}\{\gamma,\zeta\}d - \{\alpha,\gamma\}\{\gamma,\epsilon\}f}{2\gamma_1\gamma_2[\alpha,\beta][\epsilon,\zeta]}, \quad (A.3)$$

$$x_3 = \frac{1}{2\alpha_1\gamma_1\alpha_2\gamma_2[\alpha,\beta][\gamma,\delta][\zeta,\epsilon]} \times$$
$$[([\beta\delta\zeta,\alpha^2\gamma]\{\gamma,\epsilon\} + \alpha_1\alpha_2([\beta\delta,\gamma\epsilon]\{\gamma,\zeta\} + \gamma_1\gamma_2[\epsilon,\zeta]\{\beta,\delta\}))a$$
$$+ [\alpha\gamma,\beta\delta]\{\alpha,\epsilon\}\{\gamma,\epsilon\}b + \gamma_1\gamma_2[\alpha,\beta]\{\gamma,\epsilon\}(\{\alpha,\zeta\}c - \{\alpha,\epsilon\}e)$$
$$+ \alpha_1\alpha_2[\gamma,\delta]\{\alpha,\epsilon\}(\{\gamma,\zeta\}d - \{\gamma,\epsilon\}f)], \quad (A.4)$$

102   APPENDIX A. SOLVING THE ANYSTATE EQUATIONS FOR $N=3$

$$x_4 = \frac{1}{2\alpha_1\gamma_1\alpha_2\gamma_2[\gamma,\delta][\zeta,\epsilon]} \times$$
$$[\{\beta,\gamma\}(\{\alpha\gamma,\delta\zeta\}a - \{\alpha\gamma,\delta\epsilon\}b) - \gamma_1\gamma_2\{\beta,\gamma\}(\{\alpha,\zeta\}c + \{\alpha,\epsilon\}e)$$
$$+\alpha_1\alpha_2[\gamma,\delta]([\gamma,\epsilon]f - [\gamma,\zeta]d)], \quad \text{(A.5)}$$

$$x_5 = \frac{1}{2\alpha_1\gamma_1\alpha_2\gamma_2[\alpha,\beta][\zeta,\epsilon]} \times$$
$$[\{\alpha,\delta\}(\{\beta\zeta,\alpha\gamma\}a - \{\alpha\gamma,\epsilon\beta\}b) + \gamma_1\gamma_2[\alpha,\beta]([\alpha,\epsilon]e - [\alpha,\zeta]c)$$
$$+\alpha_1\alpha_2\{\alpha,\delta\}(\{\gamma,\epsilon\}f - \{\gamma,\zeta\}d)], \text{ and} \quad \text{(A.6)}$$

$$x_6 = \frac{1}{2\alpha_1\gamma_1\alpha_2\gamma_2[\alpha,\beta][\gamma,\delta][\zeta,\epsilon]} \times$$
$$[[([\alpha^2\gamma,\beta\delta\epsilon]\{\gamma,\zeta\} + \alpha_1\alpha_2([\gamma\zeta,\beta\delta]\{\gamma,\epsilon\} + \gamma_1\gamma_2[\epsilon,\zeta]\{\beta,\delta\}))b$$
$$+ \beta\delta,\alpha\gamma]\{\alpha,\zeta\}\{\gamma,\zeta\}a + \gamma_1\gamma_2[\alpha,\beta]\{\gamma,\zeta\}(\{\alpha,\zeta\}c - \{\alpha,\epsilon\}e)$$
$$+\alpha_1\alpha_2[\gamma,\delta]\{\alpha,\zeta\}(\{\gamma,\zeta\}d - \{\gamma,\epsilon\}f]. \quad \text{(A.7)}$$

In order for a solution of Eq. (A.1) to exist it is necessary that one can choose all variables from $\hat{D}_1$ and $\hat{D}_2$ such that the right hand side vector is well defined (no zero in any denominator) while at the same time its lowest two entries are zero.

Substituting the explicit forms of x_3, x_4, x_5, x_6 into Eq. (A.1), one notices that e.g. γ_1 and α_2 appear in the two zero conditions only up to first order. Hence, we solve the two equations for γ_1 and α_2, and receive along with the trivial solution ($\gamma_1 = 0, \alpha_2 = 0$):

$$\gamma_1 = \frac{\delta_1(\beta_1\zeta_1 a - \epsilon_1\beta_1 b - \alpha_1\zeta_1 d + \alpha_1\epsilon_1 f)}{\beta_1\zeta_1 c - \epsilon_1\beta_1 e - \alpha_1\zeta_1 g + \alpha_1\epsilon_1 h} \text{ and}$$
$$\alpha_2 = \frac{\beta_2(-\delta_2\zeta_2 a + \epsilon_2\delta_2 b + \gamma_2\zeta_2 c - \gamma_2\epsilon_2 e)}{-\delta_2\zeta_2 d + \epsilon_2\delta_2 f + \gamma_2\zeta_2 g - \gamma_2\epsilon_2 h}. \quad \text{(A.8)}$$

The next step is to insert these terms for γ_1 and α_2 into Eq. (A.1), more pricisely, into $\mathbf{x} = (x_1, x_2, x_3, x_4, x_5, x_6)$. Consequently, the already large terms for $\mathbf{x}$ become even larger, but we are left with a solution for $\mathbf{c} = (\alpha_3, \gamma_3, \epsilon_3, \beta_3, \delta_3, \zeta_3)$, namely $\mathbf{c} = \mathbf{x}$. Note that we are still free to choose $\alpha_1, \beta_1, \delta_1, \epsilon_1, \zeta_1$ and $\beta_2, \gamma_2, \delta_2, \epsilon_2, \zeta_2$, since they are free parameters in $\mathbf{x}$.

This freedom is only restricted by the denominators appearing in $\mathbf{x}$, as they must not become zero. After inserting the values for γ_1 and α_2 according to Eq. (A.8), these denominators are given by

$$dx_1 = 2\delta_1\beta_2[\zeta,\epsilon](\epsilon_2\delta_2 f - \delta_2\zeta_2 d + \gamma_2\zeta_2 g - \gamma_2\epsilon_2 h) \times$$
$$(\zeta_1(\beta_1\delta_2 a - \beta_1\gamma_2 c - \alpha_1\delta_2 d + \alpha_1\gamma_2 g) - \epsilon_1(\beta_1\delta_2 b - \beta_1\gamma_2 e - \alpha_1\delta_2 f + \alpha_1\gamma_2 h)), \quad \text{(A.9)}$$

$$dx_{2,4} = 2\delta_1\beta_2[\zeta,\epsilon](\beta_1\zeta_1 c - \epsilon_1\beta_1 e - \alpha_1\zeta_1 g + \alpha_1\epsilon_1 h)\times$$
$$(\beta_1(\delta_2\zeta_2 a - \epsilon_2\delta_2 b - \gamma_2\zeta_2 c + \gamma_2\epsilon_2 e) - \alpha_1(\delta_2\zeta_2 d - \epsilon_2\delta_2 f - \gamma_2\zeta_2 g + \gamma_2\epsilon_2 h)), \quad \text{(A.10)}$$

$$dx_{3,6} = 2\delta_1(\zeta_1(\beta_1\delta_2 a - \beta_1\gamma_2 c - \alpha_1\delta_2 d + \alpha_1\gamma_2 g) - \epsilon_1(\beta_1\delta_2 b - \beta_1\gamma_2 e - \alpha_1\delta_2 f + \alpha_1\gamma_2 h))\times$$
$$(\beta_1(\delta_2\zeta_2 a - \epsilon_2\delta_2 b - \gamma_2\zeta_2 c + \gamma_2\epsilon_2 e) - \alpha_1(\delta_2\zeta_2 d - \epsilon_2\delta_2 f - \gamma_2\zeta_2 g + \gamma_2\epsilon_2 h))\beta_2, \text{ and}$$
$$\text{(A.11)}$$

$$dx_5 = 2\delta_1\beta_2[\zeta,\epsilon]\times$$
$$(\beta_1(\delta_2\zeta_2 a - \epsilon_2\delta_2 b - \gamma_2\zeta_2 c + \gamma_2\epsilon_2 e) - \alpha_1(\delta_2\zeta_2 d - \epsilon_2\delta_2 f - \gamma_2\zeta_2 g + \gamma_2\epsilon_2 h)). \quad \text{(A.12)}$$

Careful examination of these terms shows that many factors reappear. Therefore, the condition that none of them be zero can be expressed in the way that the product of all different factors that appear in them must not be zero

$$\delta_1\beta_2[\zeta,\epsilon][\beta_1\zeta_1 c - \epsilon_1\beta_1 e - \alpha_1\zeta_1 g + \alpha_1\epsilon_1 h][\epsilon_2\delta_2 f - \delta_2\zeta_2 d + \gamma_2\zeta_2 g - \gamma_2\epsilon_2 h]\times$$
$$[\zeta_1(\beta_1\delta_2 a - \beta_1\gamma_2 c - \alpha_1\delta_2 d + \alpha_1\gamma_2 g) - \epsilon_1(\beta_1\delta_2 b - \beta_1\gamma_2 e - \alpha_1\delta_2 f + \alpha_1\gamma_2 h)] \times$$
$$[\beta_1(\delta_2\zeta_2 a - \epsilon_2\delta_2 b - \gamma_2\zeta_2 c + \gamma_2\epsilon_2 e) - \alpha_1(\delta_2\zeta_2 d - \epsilon_2\delta_2 f - \gamma_2\zeta_2 g + \gamma_2\epsilon_2 h)] \neq 0 \quad \text{(A.13)}$$

The remaining task is to determine the ten free parameters that were listed above in a way that fulfills the condition given in Eq. (A.13). By a proper choice of some of the remaining ten parameters this term can be simplified so that there appears only one roman letter in front of a greek letter, which means that the remaining free coefficents decouple from each other and can be chosen independently, in order to avoid factors in Eq. (A.13) to become zero. By fixing for example

$$\beta_1 = 0, \quad \delta_1 = 1, \quad \alpha_1 = 1$$
$$\beta_2 = 1, \quad \delta_2 = 0, \quad \gamma_2 = 1, \quad \text{(A.14)}$$

the condition from Eq. (A.13) becomes:

$$[\zeta,\epsilon](\zeta_1 g - \epsilon_1 h)(\zeta_2 g - \epsilon_2 h) \neq 0. \quad \text{(A.15)}$$

This equation can easily be satisfied by a proper choice of $\epsilon_1, \zeta_1, \epsilon_2, \zeta_2$, unless both, g and h, are zero. If this case should occur, one can make use of the symmetry of the state in Eq. (3.52) relating to $|+\rangle$ and $|-\rangle$, and interchange the two ground states of the atom, i. e. interchanging $|+\rangle$ and $|-\rangle$. As a consequence, one has to substitute g for b and h for a in Eq. (A.15). If in addition also a and b are zero, the formalism introduced in the next paragraph applies. So far, we have found a solution that holds for a, b, g, h

not simultaneously being zero. By constructing the solution some parameters were assigned fixed values (cf. Eq. (A.14)). With these values, γ_1 and α_2 from Eq. (A.8) become

$$\gamma_1 = \frac{\epsilon_1 f - \zeta_1 d}{\epsilon_1 h - \zeta_1 g} \quad \text{and} \quad \alpha_2 = \frac{\zeta_2 c - \epsilon_2 e}{\zeta_2 g - \epsilon_2 h}, \quad (A.16)$$

while the values of the variables of $\hat{D}_3$ are given by

$$\alpha_3 = \frac{1}{2[\zeta, \epsilon](\zeta_1 g - \epsilon_1 h)(\zeta_2 g - \epsilon_2 h)} \times$$
$$[(\zeta_2 c - \epsilon_2 e)(cd - ag)\zeta_1^2 + ((\zeta_2 cd - \epsilon_2 ed - \zeta_2 ag + \epsilon_2 bg)(\zeta_2 g - \epsilon_2 h) + \epsilon_1(\zeta_2 c - \epsilon_2 e)$$
$$(bg - de - cf + ah))\zeta_1 + (\zeta_1 g - \epsilon_1 h)(\zeta_2 g - \epsilon_2 h)(\zeta_1 c - \epsilon_1 e + \zeta_2 g - \epsilon_2 h)$$
$$+ \epsilon_1(\epsilon_1(\zeta_2 c - \epsilon_2 e)(ef - bh) + (\zeta_2 g - \epsilon_2 h)(\zeta_2(ah - cf) + \epsilon_2(ef - bh)))], \quad (A.17)$$

$$\gamma_3 = \frac{1}{2[\zeta, \epsilon](\epsilon_1 h - \zeta_1 g)(\epsilon_2 h - \zeta_2 g)} \times$$
$$[(\epsilon_2 h - \zeta_2 g)(\zeta_1 g - \epsilon_1 h)^2 + ((\zeta_2 d - \epsilon_2 f)(\epsilon_2 h - \zeta_2 g)$$
$$+ \zeta_1(\zeta_2(ag - cd) + \epsilon_2(ed - ah)) + \epsilon_1(\zeta_2(cf - bg) + \epsilon_2(bh - ef)))(\zeta_1 g - \epsilon_1 h)$$
$$- (\zeta_1 d - \epsilon_1 f)((ef - bh)\epsilon_2^2 + \zeta_2(bg - de - cf + ah)\epsilon_2 + \zeta_2^2(cd - ag))], \quad (A.18)$$

$$\epsilon_3 = \frac{1}{2(\zeta_1 g - \epsilon_1 h)(\zeta_2 g - \epsilon_2 h)} \times$$
$$[\zeta_1(cd - ag)(\epsilon_2 h - \zeta_2 g) + \epsilon_2(fg - dh)(\epsilon_2 h - \zeta_2 g) + (\zeta_1 g - \epsilon_1 h)$$
$$(\zeta_2 g^2 - \epsilon_2 hg + \epsilon_1(ch - eg)) + \epsilon_1((\epsilon_2(ah - de + cf) + \zeta_2(cd + ag))h + \epsilon_2 efg)], \quad (A.19)$$

$$\beta_3 = \frac{1}{2[\zeta, \epsilon](\zeta_1 g - \epsilon_1 h)} \times$$
$$[(cd + g(g - a))\zeta_1^2 + (\epsilon_2(fg - dh) + \epsilon_1(bg - de - cf + ah - 2gh))\zeta_1$$
$$+ \epsilon_1 \left(\zeta_2(dh - fg) + \epsilon_1 \left(h^2 - bh + ef\right)\right)], \quad (A.20)$$

$$\delta_3 = \frac{1}{2[\zeta, \epsilon](\epsilon_2 h - \zeta_2 g)} \times$$
$$[(cd + g(g - a))\zeta_2^2 + (\epsilon_1(eg - ch) + \epsilon_2(bg - de - cf + ah - 2gh))\zeta_2$$
$$+ \epsilon_2 \left(\epsilon_2 \left(h^2 - bh + ef\right) + \zeta_1(ch - eg)\right)], \text{ and} \quad (A.21)$$

$$\zeta_3 = -\frac{1}{2(\zeta_1 g - \epsilon_1 h)(\zeta_2 g - \epsilon_2 h)} \times$$
$$[(\zeta_2 g - \epsilon_2 h)(-\epsilon_1 ef + \zeta_2 gf + \epsilon_1 bh - \zeta_2 dh) + (\zeta_1 g - \epsilon_1 h)\left(\epsilon_2 h^2 - \zeta_1 ch - \zeta_2 gh + \zeta_1 eg\right)$$
$$-\zeta_1(\zeta_2 g(bg - de - cf) + \zeta_2 cdh + \epsilon_2 g(ef - bh))]. \quad (A.22)$$

The four parameters $\epsilon_{1,2}$ and $\zeta_{1,2}$ remain arbitrary with the restrictions that the condition from Eq. (A.15) be fulfilled. Note that exactly eight variables are fixed by the equations, which is exactly what one expects for a system of independent equations.

Special case: $a = b = g = h = 0$

This is the remaining case, for which the choice of parameters given in Eq. (A.14) is invalid, because denominators would be zero. To find also a solution for this case, it is again useful to write the system of equations as a pseudo-linear system of equations, but this time we extract all variables which refer to the projection of the atom a_3, i.e., the vector $\mathbf{v} = (\epsilon_1, \epsilon_2, \epsilon_3, \zeta_1, \zeta_2, \zeta_3)$:

$$\begin{pmatrix} \mathbf{M} & \mathbf{0} \\ \mathbf{0} & \mathbf{M} \end{pmatrix} \mathbf{v} = (c, d, a, g, e, f, b, h)^T \quad (A.23)$$

with $\mathbf{0}$ a 3×4 0-Matrix and

$$\mathbf{M} = \begin{pmatrix} \delta_2 \alpha_3 + \alpha_2 \delta_3 & \delta_1 \alpha_3 + \alpha_1 \delta_3 & \delta_1 \alpha_2 + \alpha_1 \delta_2 \\ \beta_2 \gamma_3 + \gamma_2 \beta_3 & \beta_1 \gamma_3 + \gamma_1 \beta_3 & \beta_1 \gamma_2 + \gamma_1 \beta_2 \\ \gamma_2 \alpha_3 + \alpha_2 \gamma_3 & \gamma_1 \alpha_3 + \alpha_1 \gamma_3 & \gamma_1 \alpha_2 + \alpha_1 \gamma_2 \\ \delta_2 \beta_3 + \beta_2 \delta_3 & \delta_1 \beta_3 + \beta_1 \delta_3 & \delta_1 \beta_2 + \beta_1 \delta_2 \end{pmatrix}. \quad (A.24)$$

Since the upper left and lower right sub-matrices are identical, we can stick in our further evaluation to a single 4×3-matrix. As we regard the case $a = b = g = h = 0$ our procedure is to choose the parameters within the matrix such that the rows belonging to a, b, g, h become linear dependent. By this, we will be left with a 6×6-matrix, whose determinant has to be chosen unequal zero, in order to obtain a unique solution for an arbitrary state.

We begin by setting

$$\begin{aligned}(\alpha_1, \alpha_2, \alpha_3) &= (\delta_1, \delta_2, \delta_3) = (1, 1, 1) \text{ and} \\ (\gamma_1, \gamma_2, \gamma_3) &= (\beta_1, \beta_2, \beta_3) = (0, \tfrac{1}{4}, \tfrac{1}{2})\end{aligned} \quad (A.25)$$

and obtain the matrix equation

$$\begin{pmatrix} 2 & 2 & 2 \\ \frac{1}{4} & 0 & 0 \\ \frac{3}{4} & \frac{1}{2} & \frac{1}{4} \\ \frac{3}{4} & \frac{1}{2} & \frac{1}{4} \end{pmatrix} \begin{pmatrix} \epsilon_1 \\ \epsilon_2 \\ \epsilon_3 \end{pmatrix} = \begin{pmatrix} c \\ d \\ 0 \\ 0 \end{pmatrix}. \qquad (A.26)$$

As already announced above, the last two rows have become linear dependent and we are left with a 3×3 -matrix, whose determinant gives $\frac{1}{8}$. The solution of (A.26) is evaluated to:

$$\epsilon_1 = 4d, \qquad \epsilon_2 = -(c + 16d), \text{ and } \quad \epsilon_3 = c + 4d. \qquad (A.27)$$

Since the lower right submatrix is identical, we obtain the same solution for the $\zeta_{1,2,3}$

$$\zeta_1 = 4f, \qquad \zeta_2 = -(e + 16f), \text{ and } \quad \zeta_3 = e + 4f. \qquad (A.28)$$

Hence, by Eq. (A.25), Eq. (A.27), and Eq. (A.28) all 18 parameters are fixed (in dependence of c, d, e, f) and by the corresponding detector setting the state

$$|\Psi\rangle = c|+-+\rangle + d|-++\rangle + e|--+\rangle + f|-+-\rangle \qquad (A.29)$$

is produced whenever all three detectors click.

Appendix B

The Natural Basis of an Arbitrary WWD

In Sec. 4.3.1, we have seen that for a certain type of WWD (two identical qubits), one finds an observable of the WWD, which allows to identify the path of the particles arriving in the minima of the interference pattern with certainty, independent of the visibility of the interference pattern. We will now tackle the question whether such an observable giving rise to a natural basis exists for all types of WWDs.

First we note that in order to find a basis in which two paths can be identified unambiguously with either of two states while a third state carries no WW information, the Hilbert space $\mathcal{H}$ of the WWD needs to have at least three dimensions. Thus, if the WWD is only a single qubit one cannot find a natural basis. However, in such a case, one may add an auxiliary qubit which does not take part in the detection process but which is also read out. In this way, $\mathcal{H}$ is enlarged to four dimensions. Therefore, we can assume that $\dim \mathcal{H} \geq 3$ always holds. In such a Hilbert space, the vectors $|\chi_a\rangle$ and $|\chi_b\rangle$ span a two-dimensional subspace (excluding the limiting case where $|\langle\chi_a|\chi_b\rangle| = 1$), and there exists thus a vector $|t\rangle \in \mathcal{H}$ which is orthogonal to both $|\chi_a\rangle$ and $|\chi_b\rangle$. In the following, we will construct an orthonormal natural basis $\{|0\rangle, |a\rangle, |b\rangle\}$ which spans the same subspace as the vectors $\{|\chi_a\rangle, |\chi_b\rangle, |t\rangle\}$.

In a first step, $|0\rangle$ is expressed as a sum of the three known vectors:

$$|0\rangle = x_1|\chi_a\rangle + x_1 e^{-i\gamma}|\chi_b\rangle + x_2|t\rangle, \tag{B.1}$$

with x_1, x_2, and γ to be determined. Since this vector is supposed to carry no WW information, the prefactors of $|\chi_a\rangle$ and $|\chi_b\rangle$ can at most differ by a relative phase γ (cf. Eq. (4.6)). In addition, we want $|\chi_a\rangle$ ($|\chi_b\rangle$) to be orthogonal to $|b\rangle$ ($|a\rangle$), because the states $|a\rangle$ and $|b\rangle$ are supposed to carry full WW information, i.e., overlap only with

APPENDIX B. THE NATURAL BASIS OF AN ARBITRARY WWD

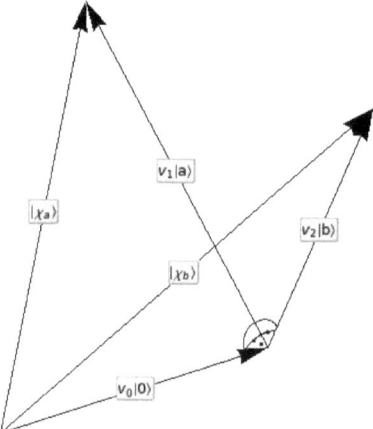

Figure B.1: The figure shows what a natural basis would look like in real vector space and it is supposed to visualize the approach for the construction of the natural basis in the complex Hilbert space. The states $|0\rangle$, $|a\rangle$, and $|b\rangle$ are pairwise orthogonal and prefactors account their non-unit length.

either $|\chi_a\rangle$ or $|\chi_b\rangle$. Thus, we can write them as

$$
\begin{aligned}
|\chi_a\rangle &= v|0\rangle + \sqrt{1-|v|^2}|a\rangle \\
|\chi_b\rangle &= ve^{i\gamma}|0\rangle + \sqrt{1-|v|^2}|b\rangle.
\end{aligned}
\tag{B.2}
$$

Fig. B.1 shows how these conditions can be fulfiled in a real threedimensional vector space. What is done in the following, is essentially to construct the corresponding vectors in the complex Hilbert space.

From Eq. (B.1), it follows immediately that the weight of $|0\rangle$ in $|\chi_a\rangle$ and $|\chi_b\rangle$ has to have the same absolute value, while the phase γ and the absolute value of v are directly defined by the overlap of $|\chi_a\rangle$ and $|\chi_b\rangle$ (remember that $|\chi_{a,b}\rangle$ are given)

$$\langle \chi_a | \chi_b \rangle = |v|^2 e^{i\gamma}. \tag{B.3}$$

Pairwise orthogonality of $|0\rangle$, $|+\rangle$, and $|-\rangle$ and the decomposition of $|\chi_a\rangle$ and $|\chi_b\rangle$ introduced in Eq. (B.2) lead to the relations

$$|\chi_a\rangle - v|0\rangle \perp |0\rangle, \tag{B.4a}$$

$$|\chi_b\rangle - ve^{i\gamma}|0\rangle \perp |0\rangle, \text{ and} \tag{B.4b}$$

$$|\chi_a\rangle - v|0\rangle \perp |\chi_b\rangle - ve^{i\gamma}|0\rangle. \tag{B.4c}$$

These conditions can be rewritten in the form of scalar products as

$$\langle \chi_a|0\rangle - v^*\langle 0|0\rangle = 0, \tag{B.5a}$$

$$\langle \chi_b|0\rangle - v^* e^{-i\gamma}\langle 0|0\rangle = 0, \text{ and} \tag{B.5b}$$

$$\langle \chi_a|\chi_b\rangle - v^*\langle 0|\chi_b\rangle - v e^{i\gamma} \underbrace{\left(\langle \chi_a|0\rangle - v^*\langle 0|0\rangle\right)}_{=0 \text{ cf. Eq. (B.5a)}} = 0. \tag{B.5c}$$

Using the notation introduced in Eq. (B.1) and Eq. (B.3) the scalar products can be expanded in the form

$$x_1 + |v|^2 x_1 - v^* = 0, \tag{B.6a}$$

$$|v|^2 x_1 + x_1 - v^* = 0, \tag{B.6b}$$

$$|v|^2 - v^*(1+|v|^2)x_1^* = 0, \text{ and} \tag{B.6c}$$

$$2(1+|v|^2)|x_1|^2 + |x_2|^2 = 1, \tag{B.6d}$$

where the first three equations are obviously pairwise equivalent, while the last equation simply denotes the normalization of $|0\rangle$. The value of x_1 can be deduced from any of the first three equations above by solving for x_1:

$$x_1 = \frac{v^*}{1+|v|^2}. \tag{B.7}$$

Since x_2 appears only as the square of an absolute value, it is determined only up to a phase by Eq. (B.6d):

$$|x_2|^2 = \frac{1-|v|^4}{(1+|v|^2)^2} \quad \Rightarrow \quad x_2 = e^{i\eta}\frac{\sqrt{1-|v|^4}}{1+|v|^2}, \tag{B.8}$$

with η arbitrary. Thus, for given $|\chi_a\rangle$ and $|\chi_b\rangle$, the vectors of the natural basis have the following form:

$$|0\rangle = \frac{v^*|\chi_a\rangle + e^{-i\gamma}v^*|\chi_b\rangle + e^{i\eta}\sqrt{1-|v|^4}|t\rangle}{1+|v|^2}, \tag{B.9a}$$

$$|a\rangle = \frac{|\chi_a\rangle - |v|^2 e^{-i\gamma}|\chi_b\rangle - v e^{i\eta}\sqrt{1-|v|^4}|t\rangle}{(1+|v|^2)\sqrt{1-|v|^2}}, \text{ and} \tag{B.9b}$$

$$|b\rangle = \frac{|\chi_b\rangle - |v|^2 e^{i\gamma}|\chi_a\rangle - v e^{i\gamma}e^{i\eta}\sqrt{1-|v|^4}|t\rangle}{(1+|v|^2)\sqrt{1-|v|^2}}, \tag{B.9c}$$

with η and the phase of v arbitrary. These three vectors are by construction pairwise orthonormal and at the same time $|a\rangle$ ($|b\rangle$) has no overlap with $|\chi_b\rangle$ ($|\chi_a\rangle$). The only condition necessary to construct these vectors was $|\langle \chi_a|\chi_b\rangle| \neq 1$ which simply states in mathematical terms that the WWD does collect some WW information. Therefore, a natural basis exists for any type of WWD, i.e., one can always choose an observable for reading out the WWD such that there exists full WW information for all particles arriving in the minima of the interference pattern, regardless of the structure of the WWD.

Bibliography

[1] A. Einstein, *Über die Erzeugung und Verwandlung des Lichtes betreffenden heuristischen Gesichtspunkt*, Annalen der Physik **17**, 132 (1905)

[2] W. Gerlach, O. Stern, *Der experimentelle Nachweis der Richtungsquantelung im Magnetfeld*, Zeitschrift für Physik **9**, 349 (1922)

[3] M. A. Nielsen, I. L. Chuang, *Quantum Computation and Quantum Information* (Cambridge University Press, Cambridge, United Kingdom, 2000)

[4] T. Tsegaye, J. Söderholm, M. Atatüre, A. Trifonov, G. Björk, A. V. Sergienko, B. E. A. Saleh, M. C. Teich, *Experimental Demonstration of Three Mutually Orthogonal Polarization States of Entangled Photons*, Physical Review Letters **85**, 5013 (2000)

[5] L. K. Shalm, R. B. A. Adamson, A. M. Steinberg, *Squeezing and over-squeezing of triphotons*, Nature **457**, 67 (2009)

[6] U. Schilling, J. von Zanthier, G. S. Agarwal, *Measuring arbitrary-order coherences: Tomography of single-mode multiphoton polarization-entangled states*, Physical Review A **81**, 013826 (2010)

[7] W. Wieczorek, R. Krischek, N. Kiesel, P. Michelberger, G. Tóth, H. Weinfurter, *Experimental Entanglement of a Six-Photon Symmetric Dicke State*, Physical Review Letters **103**, 020504 (2009)

[8] S. Bose, P. L. Knight, M. B. Plenio, V. Vedral, *Proposal for Teleportation of an Atomic State via Cavity Decay*, Phys. Rev. Lett. **83**, 24, 5158 (1999)

[9] C. Cabrillo, J. I. Cirac, P. García-Fernández, P. Zoller, *Creation of entangled states of distant atoms by interference*, Phys. Rev. A **59**, 2, 1025 (1999)

[10] A. Maser, U. Schilling, T. Bastin, E. Solano, C. Thiel, J. von Zanthier, *Generation of total angular momentum eigenstates in remote qubits*, Physical Review A **79**, 033833 (2009)

[11] U. Schilling, C. Thiel, E. Solano, T. Bastin, J. von Zanthier, *Heralded entanglement of arbitrary degree in remote qubits*, Physical Review A **80**, 022312 (2009)

[12] A. Maser, R. Wiegner, U. Schilling, C. Thiel, J. von Zanthier, *Versatile source of polarization-entangled photons*, Physical Review A **81**, 053842 (2010)

[13] W. Heisenberg, *Über den Inhalt der quantentheoretischen Mechanik und Kinematik*, Zeitschrift für Physik **43**, 172 (1927)

[14] H. P. Robertson, *The Uncertainty Principle*, Physical Review **34**, 163 (1929)

[15] N. Bohr, *Discussion with Einstein on Epistemological Problems in Atomic Physics*, pages 200–241, Library of Living Philosophers (Evanston, 1949)

[16] W. K. Wootters, W. H. Zurek, *Complementarity in the double-slit experiment: Quantum nonseperability and a quantitative statement of Bohr's principle*, Phys. Rev. D **19**, 473 (1979)

[17] G. Jaeger, A. Shimony, L. Vaidmann, *Two interferometric complementarities*, Phys. Rev. A **51**, 54 (1995)

[18] B.-G. Englert, *Fringe Visibility and Which-Way Information: An Inequality*, Phys. Rev. Lett. **77**, 2154 (1996)

[19] S. S. Afshar, E. Flores, K. F. McDonald, E. Knoesel, *Paradox in wave-particle duality*, Found. Phys. **37**, 2, 295 (2007)

[20] U. Schilling, J. von Zanthier, *et al.*, *Phase-dependent which-way information*, arXiv: quant-ph/1102.5709 (2011)

[21] U. Schilling, J. von Zanthier, *Wave-particle duality revisited*, arXiv: quant-ph/1006.2037 (2010)

[22] L. Mandel, E. Wolf, *Optical Coherence and Quantum Optics* (Cambridge University Press, Cambridge, United Kingdom, 1995)

[23] V. D'Auria, S. Fornaro, A. Porzio, S. Solimeno, S. Olivares, M. G. A. Paris, *Full Characterization of Gaussian Bipartite Entangled States by a Single Homodyne Detector*, Physical Review Letters **102**, 020502 (2009)

[24] P. G. Kwiat, K. Mattle, H. Weinfurter, A. Zeilinger, A. V. Sergienko, Y. Shih, *New High-Intensity Source of Polarization-Entangled Photon Pairs*, Phys. Rev. Lett. **75**, 4337 (1995)

[25] M. W. Mitchell, J. S. Lundeen, A. M. Steinberg, *Super-resolving phase measurements with a multiphoton entangled state*, Nature **429**, 161 (2004)

BIBLIOGRAPHY

[26] P. Walther, J.-W. Pan, M. Aspelmeyer, R. Ursin, S. Gasparoni, A. Zeilinger, *De Broglie wavelength of a non-local four-photon state*, Nature **429**, 158 (2004)

[27] C.-Y. Lu, X.-Q. Zhou, O. Gühne, W.-B. Gao, J. Zhang, Z.-S. Yuan, A. Goebel, T. Yang, J.-W. Pan, *Experimental entanglement of six photons in graph states*, Nature Physics **30**, 91 (2007)

[28] H. S. Eisenberg, G. Khoury, G. A. Durkin, C. Simon, D. Bouwmeester, *Quantum Entanglement of a Large Number of Photons*, Physical Review Letters **93**, 193901 (2004)

[29] N. Kiesel, C. Schmid, G. Tóth, E. Solano, H. Weinfurter, *Experimental Observation of Four-Photon Entangled Dicke State with High Fidelity*, Physical Review Letters **98**, 063604 (2007)

[30] T.-C. Wei, J. P. Altepeter, D. Branning, P. M. Goldbart, D. F. V. James, E. Jeffrey, P. G. Kwiat, S. Mukhopadhyay, N. A. Peters, *Synthesizing arbitrary two-photon polarization mixed states*, Physical Review A **71**, 032329 (2005)

[31] N. Mukunda, T. F. Jordan, *Determination of Optical Field Correlations from Photon Counts*, Journal of Mathematical Physics **7**, 849 (1966)

[32] G. S. Agarwal, S. Chaturvedi, *Scheme to measure quantum Stokes parameters and their fluctuations and correlations*, Journal of Modern Optics **50**, 711 (2003)

[33] Á. Rivas, A. Luis, *Practical schemes for the measurement of angular-momentum covariance matrices in quantum optics*, Physical Review A **78**, 043814 (2008)

[34] R. B. A. Adamson, P. S. Turner, M. W. Mitchell, A. M. Steinberg, *Detecting hidden differences via permutation symmetries*, Physical Review A **78**, 033832 (2008)

[35] M. Lassen, G. Leuchs, U. L. Andersen, *Continuous Variable Entanglement and Squeezing of Orbital Angular Momentum States*, Physical Review Letters **102**, 163602 (2009)

[36] M. T. L. Hsu, W. P. Bowen, P. K. Lam, *Spatial-state Stokes-operator squeezing and entanglement for optical beams*, Physical Review A **79**, 043825 (2009)

[37] R. Simon, N. Mukunda, *Minimal Three-Component SU(2) Gadget for Polarization Optics*, Physics Letters A **143**, 165 (1990)

[38] R. J. Glauber, *The Quantum Theory of Optical Coherence*, Physical Review **130**, 2529 (1963)

[39] I. N. Agafonov, M. V. Chekhova, T. S. Iskhakov, L.-A. Wu, *High-visibility intensity interference and ghost imaging with pseudo-thermal light*, Journal of Modern Optics **56**, 422 (2009)

[40] J. M. Jauch, F. Rohrlich, *The Theory of Photons and Electrons*, chapter 2.8 (Springer-Verlag, Berlin, 1980)

[41] B. A. Robson, *The Theory of Polarization Phenomena* (Clarendon Press, Oxford, 1974)

[42] M. Born, E. Wolf, *Principles of Optics*, chapter 1.4 (Cambridge University, Cambridge, 1999), 7 edition

[43] N. Korolkova, G. Leuchs, R. Loudon, T. C. Ralph, C. Silberhorn, *Polarization squeezing and continuous-variable polarization entanglement*, Physical Review A **65**, 052306 (2002)

[44] W. P. Bowen, R. Schnabel, H.-A. Bachor, P. K. Lam, *Polarization Squeezing of Continuous Variable Stokes Parameters*, Physical Review Letters **88**, 093601 (2002)

[45] E. Shchukin, W. Vogel, *Universal Measurement of Quantum Correlations of Radiation*, Physical Review Letters **96**, 200403 (2006)

[46] G. S. Agarwal, *SU(2) structure of the Poincaré sphere for light beams with orbital angular momentum*, Journal of the Optical Society of America A **16**, 2914 (1999)

[47] A. Mair, A. Vaziri, G. Weihs, A. Zeilinger, *Entanglement of the orbital angular momentum states of photons*, Nature **412**, 313 (2001)

[48] L. Allen, M. W. Beijersbergen, R. J. C. Spreeuw, J. P. Woerdman, *Orbital angular momentum of light and the transformation of Laguerre-Gaussian laser modes*, Physical Review A **45**, 8185 (1992)

[49] W. Wieczorek, C. Schmid, N. Kiesel, R. Pohlner, O. Gühne, H. Weinfurter, *Experimental Observation of an Entire Family of Four-Photon Entangled States*, Physical Review Letters **101**, 010503 (2008)

[50] N. D. Mermin, *Extreme quantum entanglement in a superposition of macroscopically distinct states*, Physical Review Letters , 65, 1838 (1990)

[51] D. M. Greenberger, M. A. Horne, A. Shimony, A. Zeilinger, *Bell's theorem without inequalities*, American Journal of Physics **58**, 1131 (1990)

[52] D. Bouwmeester, A. Ekert, A. Zeilinger, *The Physics of Quantum Information* (Springer, Berlin, 2000)

BIBLIOGRAPHY

[53] A. K. Ekert, *Quantum cryptography based on Bell's theorem*, Physical Review Letters **67**, 661 (1991)

[54] N. Gisin, G. Ribordy, W. Tittel, H. Zbinden, *Quantum cryptography*, Reviews of Modern Physics **74**, 145 (2002)

[55] C. H. Bennett, G. Brassard, C. Crépeau, R. Josza, A. Peres, W. K. Wootters, *Teleporting an unknown quantum state via dual classical and Einstein-Podolsky-Rosen channels*, Physical Review Letters **70**, 1895 (1993)

[56] R. Raussendorf, H. J. Briegel, *A One-Way Quantum Computer*, Physical Review Letters **86**, 5188 (2001)

[57] Q. A. Turchette, C. S. Wood, B. E. King, C. J. Myatt, D. Leibried, W. M. Itano, C. Monroe, D. J. Wineland, *Deterministic Entanglement of Two Trapped Ions*, Phys. Rev. Lett. **81**, 3631 (1998)

[58] E. Solano, R. L. de Matos Filho, N. Zagury, *Deterministic Bell states and measurement of the motional state of two trapped ions*, Physical Review A pages R2539–R2543, see also Physical Review A **61**, 029903(E) (2000) (1999)

[59] H. Häffner, W. Hänsel, C. F. Roos, J. Benhelm, D. Chek-al-kar, M. Chwalla, T. Körber, U. D. Rapol, M. Riebe, P. O. Schmidt, C. Becher, O. Gühne, W. Dür, R. Blatt, *Scalable multiparticle entanglement of trapped ions*, Nature **438**, 643 (2005)

[60] D. Leibfried, E. Knill, S. Seidelin, J. Britton, R. B. Blakestad, J. Chiaverini, D. B. Hume, W. M. Itano, J. D. Jost, C. Langer, R.Ozeri, R. Reichle, D. J. Wineland, *Creation of a six-atom 'Schrödinger cat' state*, Nature **438**, 639 (2005)

[61] E. Hagley, X. Maître, G. Nogues, C. Wunderlich, M. Brune, J. M. Raimond, S. Haroche, *Generation of Einstein-Podolsky-Rosen Pairs of Atoms*, Phys. Rev. Lett. **79**, 1 (1997)

[62] S.-B. Zheng, G.-C. Guo, *Efficient Scheme for Two-Atom Entanglement and Quantum Information Processing in Cavity QED*, Phys. Rev. Lett. **85**, 2392 (2000)

[63] B. Julsgaard, A. Kozhekin, E. S. Polzik, *Experimental long-lived entanglement of two macroscopic objects*, Nature **413**, 400 (2001)

[64] S. Osnaghi, P. Bertet, A. Auffeves, P. Maioli, M. Brune, J. M. Raimond, S. Haroche, *Coherent Control of an Atomic Collision in a Cavity*, Phys. Rev. Lett. **87**, 037902 (2001)

[65] D. N. Matsukevich, T. Chanelière, S. D. Jenkins, S.-Y. Lan, T. A. B. Kennedy, A. Kuzmich, *Entanglement of Remote Atomic Qubits*, Phys. Rev. Lett. **96**, 030405 (2006)

[66] D. Bouwmeester, J.-W. Pan, M. Daniell, H. Weinfurter, A. Zeilinger, *Observation of Three-Photon Greenberger-Horne-Zeilinger Entanglement*, Physical Review Letters **82**, 1345 (1999)

[67] J.-W. Pan, M. Daniell, S. Gasparoni, G. Weihs, A. Zeilinger, *Experimental Demonstration of Four-Photon Entanglement and High-Fidelity Teleportation*, Physical Review Letters **86**, 4435 (2001)

[68] M. Eibl, S. Gaertner, M. Bourennane, C. Kurtsiefer, M. Zukowski, H. Weinfurter, *Experimental Observation of Four-Photon Entanglement from Parametric Down-Conversion*, Physical Review Letters **90**, 200403 (2003)

[69] M. Bourennane, M. Eibl, C. Kurtsiefer, S. Gaertner, H. Weinfurter, O. Gühne, P. Hyllus, D. Bruß, M. Lewenstein, A. Sanpera, *Experimental Detection of Multipartite Entanglement using Witness Operators*, Physical Review Letters **92**, 087902 (2004)

[70] J.-S. Xu, C.-F. Li, G.-C. Guo, *Generation of a high-visibility four-photon entangled state and realization of a four-party quantum communication complexity scenario*, Physical Review A **74**, 052311 (2006)

[71] B. P. Lanyon, T. J. Weinhold, N. K. Langford, J. L. O'Brien, K. J. Resch, A. Gilchrist, A. G. White, *Manipulating Biphotonic Qutrits*, Physical Review Letters **100**, 060504 (2008)

[72] R. Ionicioiu, A. E. Popescu, W. J. Munro, T. P. Spiller, *Generalized parity measurements*, Physical Review A **78**, 052326 (2008)

[73] H. Hossein-Nejad, R. Stock, D. F. V. James, *Generation of multiphoton entanglement by propagation and detection*, Physical Review A **80**, 022308 (2009)

[74] B. P. Lanyon, N. K. Langford, *Experimentally generating and tuning robust entanglement between photonic qubits*, New Journal of Physics **11**, 013008 (2009)

[75] W. Wieczorek, N. Kiesel, C. Schmid, H. Weinfurter, *Multiqubit entanglement engineering via projective measurements*, Physical Review A **79**, 022311 (2009)

[76] R. Prevedel, G. Cronenberg, M. S. Tame, M. Paternostro, P. Walther, M. S. Kim, A. Zeilinger, *Experimental Realization of Dicke States of up to Six Qubits for Multiparty Quantum Networking*, Physical Review Letters **103**, 020503 (2009)

[77] L.-M. Duan, M. D. Lukin, J. I. Cirac, P. Zoller, *Long-distance quantum communication with atomic ensembles and linear optics*, Nature **414**, 413 (2001)

[78] C. Skornia, J. von Zanthier, G. S. Agarwal, E. Werner, H. Walther, *Nonclassical interference effects in the radiation from coherently driven uncorrelated atoms*, Phys. Rev. A **64**, 6, 063801 (2001)

[79] C. Simon, W. T. M. Irvine, *Robust Long-Distance Entanglement and a Loophole-Free Bell Test with Ions and Photons*, Phys. Rev. Lett. **91**, 11, 110405 (2003)

[80] L.-M. Duan, H. J. Kimble, *Efficient Engineering of Multiatom Entanglement through Single-Photon Detections*, Phys. Rev. Lett. **90**, 25, 253601 (2003)

[81] C. Thiel, J. von Zanthier, T. Bastin, E. Solano, G. S. Agarwal, *Generation of Symmetric Dicke States of Remote Qubits with Linear Optics*, Phys. Rev. Lett. **99**, 193602 (2007)

[82] T. Bastin, C. Thiel, J. von Zanthier, L. Lamata, E. Solano, G. S. Agarwal, *Operational Determination of Multiqubit Entanglement Classes via Tuning of Local Operations*, Physical Review Letters **102**, 053601 (2009)

[83] D. L. Moehring, P. Maunz, S. Olmschenk, K. C. Younge, D. N. Matsukevich, L.-M. Duan, C. Monroe, *Entanglement of single-atom quantum bits at a distance*, Nature **449**, 68 (2007)

[84] X.-L. Feng, Z.-M. Zhang, X.-D. Li, S.-Q. Gong, Z.-Z. Xu, *Entangling Distant Atoms by Interference of Polarized Photons*, Physical Review Letters **90**, 217902 (2003)

[85] G. S. Agarwal, *Quantum Optics, Vol. 70 of Springer Tracts in Modern Physics* (Springer-Verlag, Berlin, 1974)

[86] B. B. Blinov, D. L. Moehring, L.-M. Duan, C. Monroe, *Observation of entanglement between a single trapped atom and a single photon*, Nature **428**, 153 (2004)

[87] J. Volz, M. Weber, D. Schlenk, W. Rosenfeld, J. Vrana, K. Saucke, C. Kurtsiefer, H. Weinfurter, *Observation of Entanglement of a Single Photon with a Trapped Atom*, Phys. Rev. Lett. **96**, 030404 (2006)

[88] S. Gerber, D. Rotter, M. Hennrich, R. Blatt, F. Rohde, C. Schuck, M. Almendros, R. Gehr, F. Dubin, J. Eschner, *Quantum interference from remotely trapped ions*, New Journal of Physics **11**, 013032 (2009)

[89] U. Eichmann, J. C. Bergquist, J. J. Bollinger, J. M. Gilligan, W. M. Itano, D. J. Wineland, M. G. Raizen, *Young's Interference Experiment with Light Scattered from Two Atoms*, Physical Review Letters **70**, 2359 (1993)

[90] G. S. Agarwal, J. von Zanthier, C. Skornia, H. Walther, *Intensity-intensity correlations as a probe of interferences und conditions of noninterference in the intensity*, Physical Review A **65**, 053826 (2002)

[91] S. Gerber, D. Rotter, M. Hennrich, R. Blatt, F. Rohde, C. Schuck, M. Almendros, R. Gehr, F. Dubin, J. Eschner, *Title: Quantum interference from remotely trapped ions*, arXiv: quant-ph/0810.1847 (2008)

[92] R. Horodecki, P. Horodecki, M. Horodecki, K. Horodecki, *Quantum entanglement*, Reviews of Modern Physics **81**, 865 (2009)

[93] W. K. Wootters, *Entanglement of Formation of an Arbitrary State of Two Qubits*, Physical Review Letters **80**, 2245 (1998)

[94] E. Hecht, *Optics* (Addison Wesley, San Francisco, California, USA, 2002)

[95] F. Tamburini, B. A. Bassett, C. Ungarelli, *Detecting gravitational waves using entangled photon states*, Phys. Rev. A **78**, 012114 (2008)

[96] W. Dür, G. Vidal, J. I. Cirac, *Three qubits can be entangled in two inequivalent ways*, Physical Review A **62**, 062314 (2000)

[97] L. Lamata, J. León, D. Salgado, E. Solano, *Inductive entanglement classification of four qubits under stochastic local operations and classical communication*, Physical Review A **75**, 022318 (2007)

[98] T. Bastin, S. Krins, P. Mathonet, M. Godefroid, L. Lamata, E. Solano, *Operational Families of Entanglement Classes for Symmetric N-Qubit States*, Physical Review Letters **103**, 070503 (2009)

[99] R. H. Dicke, *Coherence in Spontaneous Radiation Processes*, Phys. Rev. **93**, 99 (1954)

[100] Hagiwara, *et al.*, *Review of Particle Properties*, Physical Review D **66**, 010001 (2002)

[101] C. Ammon, *Generation of total angular momentum eigenstates in remote qubits: A Proof*, Master's thesis, Erlangen-Nürnberg (2010)

[102] O. Mandel, M. Greiner, A. Widera, T. Rom, T. W. Hänsch, I. Bloch, *Controlled collisions for multi-particle entanglement of optically trapped atoms*, Nature **425**, 937 (2003)

[103] M. A. Nielsen, *Optical Quantum Computation Using Cluster States*, Physical Review Letters **93**, 040503 (2004)

[104] Y. L. Lim, A. Beige, L. C. Kwek, *Repeat-Until-Success Linear Optics Distributed Quantum Computing*, Physical Review Letters **95**, 030505 (2005)

[105] M. Borhani, D. Loss, *Cluster states from Heisenberg interactions*, Physical Review A **71**, 034308 (2005)

[106] N. Kiesel, C. Schmid, U. Weber, G. Tóth, O. Gühne, R. Ursin, H. Weinfurter, *Experimental Analysis of a Four-Qubit Photon Cluster State*, Physical Review Letters **95**, 210502 (2005)

[107] S. D. Barrett, P. Kok, *Efficient high-fidelity quantum computation using matter qubits and linear optics*, Physical Review A **71**, 060310(R) (2005)

[108] D. E. Browne, T. Rudolph, *Resource-Efficient Linear Optical Quantum Computation*, Physical Review Letters **95**, 010501 (2005)

[109] Y. S. Weinstein, C. S. Hellberg, J. Levy, *Quantum-dot cluster-state computing with encoded qubits*, Physical Review A **72**, 020304(R) (2005)

[110] P. Walther, K. J. Resch, T. Rudolph, E. Schenck, H. Weinfurter, V. Vedral, M. Aspelmeyer, A. Zeilinger, *Experimental one-way quantum computing*, Nature **434**, 169 (2005)

[111] S. C. Benjamin, D. E. Browne, J. Fitzsimons, J. J. L. Morton, *Brokered graph-state quantum computation*, New Journal of Physics **8**, 141 (2006)

[112] T. P. Bodiya, L.-M. Duan, *Scalable Generation of Graph-State Entanglement Through Realistic Linear Optics*, Physical Review Letters **97**, 143601 (2006)

[113] L.-M. Duan, M. J. Madsen, D. L. Moehring, P. Maunz, R. N. Kohn, C. Monroe, *Probabilistic quantum gates between remote atoms through interference of optical frequency qubits*, Physical Review A **73**, 062324 (2006)

[114] T. Tanamoto, Y.-X. Liu, S. Fujita, X. Hu, F. Nori, *Producing Cluster States in Charge Qubits and Flux Qubits*, Physical Review Letters **97**, 230501 (2006)

[115] K. Chen, C.-M. Li, Q. Zhang, Y.-A. Chen, A. Goebel, S. Chen, A. Mair, J.-W. Pan, *Experimental Realization of One-Way Quantum Computing with Two-Photon Four-Qubit Cluster States*, Physical Review Letters **99**, 120503 (2007)

[116] R. Prevedel, P. Walther, F. Tiefenbacher, P. Böhi, R. Kaltenbaek, T. Jennewein, A. Zeilinger, *High-speed linear optics quantum computing using active feed-forward*, Nature **445**, 65 (2007)

[117] M. S. Tame, M. Paternostro, M. S. Kim, *One-way quantum computing in a decoherence-free subspace*, New Journal of Physics **9**, 201 (2007)

[118] G. Vallone, E. Pomarico, F. De Martini, P. Mataloni, *Active One-Way Quantum Computation with Two-Photon Four-Qubit Cluster States*, Physical Review Letters **100**, 160502 (2008)

[119] B. Vaucher, A. Nunnenkamp, D. Jaksch, *Creation of resilient entangled states and a resource for measurement-based quantum computation with optical superlattices*, New Journal of Physics **10**, 023005 (2008)

[120] Y. Xia, J. Song, A.-D. Zhu, Z. Jin, S. Zhang, H.-S. Song, *Preparation of a class of multiatom entangled states*, Journal of the Optical Society of America B **26**, 1599 (2009)

[121] Q.-B. Fan, L. Zhou, *Generation of Cluster-Type Entangled Coherent States via Cavity QED*, International Journal of Theoretical Physics **49**, 128 (2010)

[122] Q. Lin, B. He, *Efficient generation of universal two-dimensional cluster states with hybrid systems*, Physical Review A **82**, 022331 (2010)

[123] F.-Y. Zhang, P. Pei, C. Li, H.-S. Song, *One-step generation of cluster states in multiple flux qubits coupled with a nanomechanical resonator*, Physica B **405**, 3334 (2010)

[124] R. Raussendorf, D. E. Browne, H. J. Briegel, *Measurement-based quantum computation on cluster states*, Physical Review A **68**, 022312 (2003)

[125] S. C. Benjamin, J. Eisert, T. M. Stace, *Optical generation of matter qubit graph states*, New Journal of Physics **7**, 194 (2005)

[126] S. Krinner, *Generation of a Three-Qubit Anystate*, Technical report, Friedrich-Alexander-Universität Erlangen-Nürnberg, project work within in the programme "Physik mit integriertem Doktorandenkolleg" (2009)

[127] B. Buchberger, in *Progress, directions and open problems in multidimensional systems theory*, pages 184–232 (Reidel, 1985)

[128] W. Boege, R. Gebauer, H. Kredel, *Some Examples for Solving Systems of Algebraic Equations by Calculating Groebner Bases*, Journal of Symbolic Computation **1**, 83 (1986)

[129] F. Verstraete, J. Dehaene, B. De Moor, H. Verschelde, *Four qubits can be entangled in nine different ways*, Physical Review A **65**, 052112 (2002)

[130] L. Lamata, J. León, D. Salgado, E. Solano, *Inductive classification of multipartite entanglement under stochastic local operations and classical communicationLe*, Physical Review A **74**, 052336 (2006)

[131] M. Aspelmeyer, J. Eisert, *Quantum mechanics: Entangled families*, Nature **455**, 180 (2008)

[132] N. Kiesel, W. Wieczorek, S. Krins, T. Bastin, H. Weinfurter, E. Solano, *Operational multipartite entanglement classes for symmetric photonic qubit states*, Physical Review A **81**, 032316 (2010)

[133] T. Bastin, S. Krins, P. Mathonet, M. Godefroid, L. Lamata, E. Solano, *Operational Families of Entanglement Classes for Symmetric N-Qubit States*, Physical Review Letters **103**, 070503 (2009)

[134] P. Mathonet, S. Krins, M. Godefroid, L. Lamata, E. Solano, T. Bastin, *Entanglement equivalence of N-qubit symmetric states*, Physical Review A **81**, 052315 (2010)

[135] C. Held, *Die Bohr-Einstein-Debatte, Quantenmechanik und physikalische Wirklichkeit* (Mentis-Verlag, Paderborn, 1998)

[136] T. Young, *Experimental Demonstration of the General Law of the Interference of Light*, Philosophical Transactions of the Royal Society of London **92**, 12 (1802)

[137] M. O. Scully, B.-G. Englert, H. Walther, *Quantum optical tests of complementarity*, Nature **351**, 111 (1991)

[138] L. Mandel, *Coherence and indistinguishability*, Opt. Lett. **16**, 1882 (1991)

[139] S. Dürr, T. Nonn, G. Rempe, *Origin of quantum-mechanical complementarity probed by a which-way experiment in an atom interferometer*, Nature **395**, 33 (1998)

[140] E. Buks, R. Schuster, M. Heiblum, D. Mahalu, V. Umansky, *Dephasing in electron interference by a 'which-path' detector*, Nature **391**, 871 (1998)

[141] G. Björk, A. Karlsson, *Complementarity and quantum erasure in welcher Weg experiments*, Phys. Rev. A **58**, 3477 (1998)

[142] G. J. Pryde, J. L. O'Brien, A. G. White, S. D. Bartlett, T. C. Ralph, *Measuring a Photonic Qubit without Destroying It*, Phys. Rev. Lett. **92**, 19, 190402 (2004)

[143] X. Peng, X. Zhu, D. Suter, J. Du, M. Liu, K. Gao, *Quantification of complementarity in multiqubit systems*, Phys. Rev. A **72**, 052109 (2005)

[144] M. Kolář, T. Opatrný, N. Bar-Gill, N. Erez, G. Kurizki, *Path–phase duality with intraparticle translational–internal entanglement*, New J. Phys. **9**, 129 (2007)

[145] V. Jacques, E. Wu, F. Grosshans, F. Treussart, P. Grangier, A. Aspect, J.-F. Roch, *Delayed-Choice Test of Quantum Complementarity with Interfering Single Photons*, Phys. Rev. Lett. **100**, 220402 (2008)

[146] M. Barbieri, M. E. Goggin, M. P. Almeida, B. P. Lanyon, A. G. White, *Complementarity in variable strength quantum non-demolition measurements*, New J. Phys. **11**, 093012 (2009)

[147] N. Erez, D. Jacobs, G. Kurizki, *Operational path-phase complementarity in single-photon interferometry*, J. Phys. B **42**, 114006 (2009)

[148] M. O. Scully, K. Drühl, *Quantum eraser: A proposed photon correlation experiment concerning observation and "delayed choice" in quantum mechanics*, Phys. Rev. A **25**, 2208 (1982)

[149] M. Hillery, M. O. Scully, *Quantum Optics, Experimental Gravity, and Measurement Theory*, page 65 (Plenum, New York, 1983)

[150] P. G. Kwiat, A. M. Steinberg, R. Y. Chiao, Phys. Rev. A **45**, 7729 (1992)

[151] T. J. Herzog, P. G. Kwiat, H. Weinfurter, A. Zeilinger, *Complementarity and the Quantum Eraser*, Phys. Rev. Lett. **75**, 17, 3034 (1995)

[152] Y.-H. Kim, R. Yu, S. P. Kulik, J. Shih, M. O. Scully, *Delayed "Choice" Quantum Eraser*, Phys. Rev. Lett. **84**, 1 (2000)

[153] B.-G. Englert, J. A. Bergou, *Quantitative quantum erasure*, Opt. Comm. **179**, 337 (2000)

[154] J.-M. Lévy-Leblond, *Quantum Physics and Language*, Physica **151B**, 314 (1988)

[155] B.-G. Englert, *Remarks on some basic issues in quantum mechanics*, Z. Naturforsch. **54a**, 11 (1999)

[156] B.-G. Englert, M. O. Scully, H. Walther, *Quantum erasure in double-slit interferometers with which-way detectors*, Am. J. Phys. **67**, 325 (1999)

Danksagung

Mein erster Dank geht an meinem Doktorvater *Prof. Dr. Joachim von Zanthier*, der mir die Bearbeitung dieses spannenden Themas ermöglichte. Seine stete Bereitschaft zur enthusiastischen Diskussion neuer Ergebnisse und seine fortwährende Unterstützung waren die Grundlage für das Entstehen dieser Arbeit.

Mein Dank gilt auch *Prof. Girish S. Agarwal, FRS, DSc. (h.c.)*, für seine Einladung an die Oklahoma State University, Stillwater, OK, USA. Während der drei Monate meines Aufenthalts entstand der erste Teil der vorliegenden Arbeit, und auch während seiner Besuche in Erlangen war immer Zeit, aktuelle Ergebnisse und grundlegende Fragen zu erörtern.

Des Weiteren danke ich *Prof. Dr. Thierry Bastin* und *Prof. Dr. Enrique Solano* für ihre Bereitschaft zur Kollaboration und ihre engagierte Zusammenarbeit.

Ich freue mich über die finanzielle Unterstützung des Elitenetzwerks Bayern, die mir möglich machte, mich voll auf meine Forschungsarbeit zu konzentrieren.

Ein herzliches Dankeschön auch an Sebastian und Jason, die im Rahmen des beschleunigten Studiengangs für Physik kleinere Projekte auf meinem Forschungsgebiet bearbeiteten.

Meinen Mitdoktoranden Christoph, Steffen, Ralph und Andreas, sowie den anderen zeitweiligen Mitgliedern unserer Gruppe danke ich für die freundschaftliche Atmosphäre und interessanten fachlichen und außerfachlichen Austausch.

Und schließlich möchte ich mich noch bei meiner Familie und meinen Freunden bedanken, bei meinen Eltern für die Unterstützung auf meinem bisherigen Weg, bei meiner Freundin Romy für das Verständnis und die Liebe während einer nicht immer ganz einfachen Zeit, und bei meinem Sohn Phileas, für die Zerstreuung und Ablenkung, die seine Anwesenheit mir gebracht haben.

Die VDM Verlagsservicegesellschaft sucht für wissenschaftliche Verlage abgeschlossene und herausragende

Dissertationen, Habilitationen, Diplomarbeiten, Master Theses, Magisterarbeiten usw.

für die kostenlose Publikation als Fachbuch.

Sie verfügen über eine Arbeit, die hohen inhaltlichen und formalen Ansprüchen genügt, und haben Interesse an einer honorarvergüteten Publikation?

Dann senden Sie bitte erste Informationen über sich und Ihre Arbeit per Email an *info@vdm-vsg.de*.

Sie erhalten kurzfristig unser Feedback!

VDM Verlagsservicegesellschaft mbH
Dudweiler Landstr. 99   Telefon +49 681 3720 174
D - 66123 Saarbrücken   Fax   +49 681 3720 1749
www.vdm-vsg.de

Die VDM Verlagsservicegesellschaft mbH vertritt

Printed by Books on Demand GmbH, Norderstedt / Germany